Aquaculture: Principles and Practice

Aquaculture: Principles and Practice

Summer Walter

New York

Published by Syrawood Publishing House,
750 Third Avenue, 9th Floor,
New York, NY 10017, USA
www.syrawoodpublishinghouse.com

Aquaculture: Principles and Practice
Summer Walter

© 2021 Syrawood Publishing House

International Standard Book Number: 978-1-64740-071-2 (Hardback)

This book contains information obtained from authentic and highly regarded sources. All chapters are published with permission under the Creative Commons Attribution Share Alike License or equivalent. A wide variety of references are listed. Permissions and sources are indicated; for detailed attributions, please refer to the permissions page. Reasonable efforts have been made to publish reliable data and information, but the authors, editors and publisher cannot assume any responsibility for the validity of all materials or the consequences of their use.

Trademark Notice: Registered trademark of products or corporate names are used only for explanation and identification without intent to infringe.

Cataloging-in-Publication Data

Aquaculture : principles and practice / Summer Walter.
p. cm.
Includes bibliographical references and index.
ISBN 978-1-64740-071-2
1. Aquaculture. 2. Agriculture. I. Walter, Summer.
SH21 .A68 2021
639.8--dc23

TABLE OF CONTENTS

PREFACE

The purpose of this book is to help students understand the fundamental concepts of this discipline. It is designed to motivate students to learn and prosper. I am grateful for the support of my colleagues. I would also like to acknowledge the encouragement of my family.

Aquaculture refers to the farming of aquatic organisms including fish, molluscs, aquatic plants, algae and other organisms. It primarily deals with the cultivation of freshwater and saltwater populations of these organisms. There are various kinds of aquaculture such as fish farming, algaculture, oyster farming, shrimp farming and mariculture. Fish farming is one of the most widely practiced farming in aquaculture that involves raising fish for commercial purposes. The most important species of fish produced in fish farming are salmon, catfish, tilapia and carp. Algaculture involves the farming of different species of algae. The common methods of aquaculture are aquaponics and integrated multi-trophic aquaculture. The topics included in this book on aquaculture and fish farming are of utmost significance and bound to provide incredible insights to readers. Most of the topics introduced herein cover new techniques and the applications of aquaculture and fish farming. This book will serve as a valuable source of reference for those interested in this field.

A foreword for all the chapters is provided below:

Chapter - What is Aquaculture?

Aquaculture is defined as the farming of aquatic plants, algae and various aquatic animals such as fish, molluscs, crustaceans, etc. This is an introductory chapter which will introduce briefly all the significant aspects of aquaculture as well as its purpose and benefits.

Chapter - Branches of Aquaculture

There are a numerous branches of aquaculture such as fish farming, shrimp farming, oyster farming, mariculture, offshore aquaculture, coral aquaculture, inland saline aquaculture and algaculture. The topics elaborated in this chapter will help in gaining a better perspective about these branches of aquaculture.

Chapter - Species Farmed in Aquaculture

There are various species which are bred, reared and harvested in aquaculture. The common species dealt with in aquaculture are aquatic plants, fish, crustaceans, molluscs, echinoderm, jellyfish, etc. The topics elaborated in this chapter will help in gaining a better perspective about these species.

Chapter - Methods and Techniques

The numerous techniques used in aquaculture include raceway, aquaponics, recirculating aquaculture systems, etc. All these methods and techniques related to aquaculture as well as practices such as using antimicrobials and copper alloys has been carefully analyzed in this chapter.

Chapter - Fish Farming

The practice of raising fish in tanks or enclosures primarily for food is known as fish farming. Some of the commonly farmed fish are catfish, cobia, tilapia and salmonids. This chapter discusses in detail the aquaculture of these fish as well as the various aspects of fish farming such as fish hatchery and fish stocking.

Chapter - Algaculture

Algaculture is a branch of aquaculture that deals with the farming of algae species. They are broadly divided into two main categories, namely, microalgae and macroalgae. The diverse aspects of algaculture as well as the diverse uses for algae have been thoroughly discussed in this chapter.

Summer Walter

What is Aquaculture? 1

Aquaculture is defined as the farming of aquatic plants, algae and various aquatic animals such as fish, molluscs, crustaceans, etc. This is an introductory chapter which will introduce briefly all the significant aspects of aquaculture as well as its purpose and benefits.

Aquaculture is the science, art, or practice of cultivating and harvesting aquatic organisms, including fish, mollusks, crustaceans, aquatic plants, and algae such as seaweed. Operating in marine, brackish, and freshwater environments, aquaculture provides food for people and in smaller amounts supplies fish for stocking lakes, bait for fishing, and live specimens for home aquariums. Aquacultural practices span a great range from poor farmers with a few carp in a pond in China to commercial growers raising millions of shrimp on the mangrove coasts of Indonesia or millions of salmon in net cages in Norwegian fiords.

Mariculture, or marine aquaculture, is the subset of aquaculture specifically related to all forms of marine organisms, including finfish, shellfish and marine plants. The term "fish farming" is often used synonymously with aquaculture. In popular use, the term fish, when used for food, generally conveys more than the taxonomic grouping of fish, extending to invertebrates such as lobsters, crabs, shellfish, and so forth. The term "fishery" has similar broad applications. Some individuals, however, consider fish farming as a subset of aquaculture, limiting it to fish or to particular technologies.

China is by far the largest producer of aquacultural products, accounting in 2004 for nearly 70 percent of the worldwide aquacultural output, with the rest of the Asia and Pacific region accounting for another 22 percent of the production, and the remaining 8 percent being distributed between Europe, Africa, the Middle East, and North and

South America. In China, the cyprinids (including carp) are the dominant farmed fish, while in Western Europe the salmonids (including Atlantic salmon) hold that position, and in North America the dominant aquacultural product is the catfish.

Aquaculture is in a period of rapid expansion as diverse people and nations turn to it as a means of alleviating world hunger and providing food and a cash crop for families. Aquaculture offers the promise of fostering a food production system that is ecologically sustainable and able to alleviate stresses on wild populations of aquatic organisms; but realizing that promise remains difficult.

Whatever the context, the rapid expansion of aquacultural activities tends to follow a first course of pursuing the greatest productivity at the lowest cost while relying on the free use of such environmental goods and services as wild fish (to make fish meal) and water currents (to flush away wastes). Practiced in this way, aquaculture, which is a critical component of the global food supply system, may be neither sustainable nor a benefit to wild populations. Fortunately considerable progress in developing and applying sustainable aquacultural techniques has already been made and much that has been learned is transferable. Efforts to move commercial aquacultural facilities toward sustainability are being undertaken both by regulatory agencies and by the growers themselves. The practice of aquaculture is ancient and found in many cultures.

Aquaculture was used in China circa 3500 B.C.E. When the waters lowered after river floods, some fishes, namely carp, were held in artificial lakes. Their brood were later fed using nymphs and feces from silkworms used for silk production. The oldest known document on fish culture was written by a Chinese politician, Fan-Li, in 475 B.C.E.

The Hawaiian people practiced aquaculture by constructing fish ponds, with an organized system in place by 400 C.E. A remarkable example from ancient Hawaii is the Menehune fish pond thought to date from at least one thousand years ago, at Alekoko on Kauai. According to legend, it was constructed in one night by the little people called Menehune.

Egypt, Europe and the Americas also have a long history of aquaculture. There is evidence of Egyptian aquaculture, focusing on tilapia, tracing to 2000 B.C.E., while Roman aquaculture was practiced in the first century B.C.E., according to Pliny the Elder. The Romans focused on trout and mullet and were quite adept at breeding fish in ponds. In Central Europe, there is record of pond fish culture at the end of the eleventh century. In Europe during the Middle Ages, aquaculture became common in monasteries, as fish was scarce and thus expensive. A fourteenth century French monk, Dom Pinchon, may have been the first person to artificially fertilize trout eggs. There is also evidence that the Maya practiced a form of aquaculture, as did the native peoples of North America.

Transportation improvements in the nineteenth century made wild fish caught by fishermen easily available and inexpensive, even far from the sea, causing a decline in aquaculture.

If China is considered the cradle of aquaculture, then perhaps France is the birthplace of modern aquaculture, with the first fish hatchery having been established there in 1852. The current boom in aquaculture started in the 1960s as prices for fish began to climb. Wild fish capture was reaching its peak and the human population was continuing to rise. Today, commercial aquaculture exists on an unprecedented, massive scale.

In the 1980s, open-net cage salmon farming was also expanding; this particular type of aquaculture technology is still a minor part of the production of farmed finfish worldwide. However, evidence of its negative impact on wild stocks, which started coming to light in the late 1990s, has caused it to be a major cause of controversy.

With the global human population increasing steadily and hunger confronting millions of people, food production is a pressing concern calling ideally for high quality, nutritious food in large quantities from a source that does not severely disrupt the environment.

Many people have held out the hope that aquaculture can be one such source. Indeed, aquaculture has been one of the fastest growing segments of the global food production sector in recent decades. By 2004, aquaculture had grown to be a US$70 billion industry contributing almost one-half of the fish products consumed by humans. About one-half of the aquaculture output was in the form of finfish. Algae and aquatic plants made up almost one-quarter of the production by weight, and mollusks (including oysters, clams, scallops, and mussels) about one-fifth by weight.

Aquaculture has generated considerable interest because fish and other aquatic organisms are generally very efficient converters of feedstuffs into high quality protein when compared to other farmed animals. For example, a catfish may require 6 kg of feed (wet weight to wet weight) to produce 1 kg of catfish whereas a chicken might require 10 kg and a pig 30 kg. This is possible primarily because aquatic species are cold-blooded (or more correctly, poikilothermic), and hence do not expend energy on heating, and because movement in the aquatic environment requires little energy. Fish and other aquatic organisms also tend to have a higher percentage of edible weight than terrestrial species.

Aquaculture can produce large quantities of protein in a relatively small space, and recent developments in technology mean that the water can be filtered and re-used, providing more efficient use of water resources than the simple flow through systems. Furthermore, aquaculture can be easily integrated with other crops, particularly vegetable crops. In China and southeast Asia, rice farmers raise carp in their rice paddies. Outside rice growing areas, the fish wastewater, rich in nutrients, is excellent fertilizer that can be used for irrigation or as the nutrient-rich water for growing vegetables hydroponically.

Furthermore, farming of high value (and often overexploited) species can reduce pressure on wild stocks, and even help in the replenishing of wild stocks.

Types of Growing Systems

Pond Culture

The simplest system for raising fish is in ponds or irrigation ditches. Juvenile fish or fingerlings are put into a pond and fed until they reach market size. The fish are caught, either by draining the pond or by using large nets. Food can be from natural sources—commonly zooplankton feeding on pelagic algae, or benthic animals, such crustaceans and mollusks. Tilapia species feed directly on phytoplankton, making higher production possible.

There are a number of factors that determine the amount of fish that any given pond can produce. The first is the size of the pond, which determines the amount of water available for the fish, which in turn determines the amount of oxygen available for the fish. If there are too many fish in the pond, there will not be enough oxygen, and the fish will become stressed and begin to die. Another factor is the capacity of the pond to digest waste from the fish and the uneaten feed. The waste that is toxic to fish is mostly in the form of ammonia, nitrites, and nitrates.

The pond environment provides natural ways to eliminate waste. For example, in one waste processing cascade, the initiating bacteria convert available ammonia to available nitrites, which a second bacteria converts to the available nitrates that plants and algae consume as a growth nutrient. The viable density of fish in a pond is determined by the balance between the amount of waste generated and natural processes for waste elimination. If the fish release too much waste into the pond, the natural processes cannot keep up and the fish will become stressed.

Fish density can be increased if fresh water can be introduced to the pond to flush out wastes or if the pond can be aerated, either with compressed air or mechanically by using paddle wheels. Adding oxygen to the water not only increases the amount of oxygen in the water available for the fish, it also improves the processes involved in removing the wastes.

Another factor affecting pond culture is predation from birds such as egrets and herons, and animals such as raccoons, otters, and even bears in some areas. If the pond is small, fences and overhead netting can control predation. When ponds are large, however, predation is very problematic. In some cases, farms have been ruined by bird predation.

Another concern is algal blooms, which can lead to an exhaustion of nutrients, followed by a die-off of the algae, depletion of the oxygen, and pollution of the water, leading to a loss of fish.

Advantages of pond culture include its simplicity, and relatively low labor requirements (apart from the harvesting of the fish). It also has low energy requirements. A major disadvantage is that the farm operation is more dependent on weather and other natural factors that are beyond the farmer's control. Another disadvantage concerns the marketing of the fish. Generally, ponds are only harvested when most of the fish are at market size. This means the farmer has many fish to market at the same time, requiring a market that can absorb large quantities of fish at a time and still give a good price to the farmer. Usually this means there is a need for some kind of processing and large-scale marketing, with several fish farms in the same area to provide the processing plant with a constant supply of fish. If this kind of marketing infrastructure is not available, then it is difficult for the fish farmer.

Cage Culture

Cage-based aquaculture cultivates aquatic organisms by confining them in a cage within a body of water, which could be a pond, a river, or an estuary.

In cage culture, the waste from the organisms and food they don't eat is passed to the receiving body of water with the expectation that natural processes will remove the waste from the water. In such systems, the grower needs achieve a balance between the density of aquatic organisms in each cage and the number of cages in the body of water and the amount of waste the body of water can absorb and still maintain acceptable water quality. The problem of pollution always occurs when too many fish are put in too little water, whether it is in a pond or a cage or several cages placed together in a larger water body. So long as the proper balance is maintained, however, pollution is not a problem. The farmer must then determine if that balance will provide enough production to be profitable.

An advantage of the cage culture is that the farmer has more control over the fish and multiple harvests are possible with less labor. This allows for more marketing options when smaller quantities of fish are harvested over longer periods of time. For example, the farmer may be able to market his fish to local restaurants or fresh fish markets and therefore be able to command a better price for his fish. Another advantage is that the cages generally provide protection from most predators.

The major disadvantage of pond culture is that the amount of fish the farm can produce is limited by the amount of waste the receiving water can absorb and digest. This in turn limits the growth potential of the farm. Another disadvantage is that the cages are vulnerable to storms, floods, and in some cases, winter ice.

Flow-through System

A flow-through system involves the movement of water through a series of raceways or tanks. The waste is flushed out of the system into a receiving body of water. In many cases, the raceways or tanks may simply be downstream of a diverted river or stream. This can be a simple system if there is a sufficient amount of clean water entering the system.

The raceways or tanks give better control of the feeding and allow for better handling of the fish. It also is easier to grade or sort the fish by size. As with the cage system, the raceways make it possible to harvest and market fish over a longer period of time, which improves the farmers marketing position. Some farms have pumped well water instead of using streams or springs. The advantage of pumping from wells is that water purity can be controlled more easily and there is less chance for disease and parasites contaminating the fish.

The major disadvantage of the flow-through system is the cost of pumping the large amount of water that is required. The optimum fish density for this system is limited by the flow rate of clean water and the amount of waste the receiving water can absorb while maintaining acceptable water quality.

Recirculating Systems

Two problems common to the pond, cage, and flow-through systems are that they all require a large amount of clean water, and the environment must absorb a large amount of waste. Since the 1960s, much research and experimentation has been done on recirculating systems.

A recirculating system means that the aquatic organisms are grown in raceways or tanks and the waste is then removed from the water and the water re-used. Oxygen is added to the water at a constant rate by using compressed air or liquid oxygen, or via cascade flow. The recycling system uses natural processes to remove the waste, but confines and intensifies the processes in order to get more waste removed in less time and using less space. Mechanical filters such as settling tanks, pad filters, or rotating screens called drum filters remove the solid waste. Another type of filter is the biological filter. This filter removes ammonia and nitrite from the water, which come from the fish body waste. These two chemicals are particularly toxic to fish. The filter uses bacteria to digest the ammonia and nitrite, with the bacteria growing on surfaces inside the filter. The surface area is critical to the efficiency of the filter. The more surface area, the more bacteria and the more bacteria, the more waste that can be removed. The surface area is usually shredded plastic, plastic rings, or plastic beads. Plastic is usually used because it is durable and can be cleaned and re-used.

The major advantage in this system is that large numbers of aquatic organisms can be raised in very little space and using a small amount of water. Another advantage is that the farmer can manage the organisms, the water quality, the water temperature, the feed rations, and the grading much more closely, especially if the tanks are in a building. This allows the farmer to plan for year round production, which is a strong marketing advantage. It also solves the predation problem and weather related problems. The problem of waste removal is easier to manage with a closed system. The waste can be totally removed from the system and spread on land as fertilizer, rather than dumped into a receiving body of water.

A major disadvantage to recirculation systems is the capital cost to construct the system. A system that can produce 100,000 pounds of fish a year can cost up to US$500,000, not including the land or buildings. Another major problem is the energy cost to run the system, as well as the higher cost of fish food, which must contain a much higher level of protein (up to 60 percent) than, for example, cattle food, and a balanced amino acid composition as well.

A third area of concern is disease and parasites. When fish densities are high, the risk of infections by parasites like fish lice; fungi (Saprolegnia ssp.; intestinal worms, such as nematodes or trematodes; bacteria (e.g., Yersinia ssp, Pseudomonas ssp.), and protozoa (such as Dinoflagellates) is much higher than in animal husbandry because of the ease in which pathogens can invade the fish body (e.g. by the gills). Once a system is infected, it is very difficult to purge the system. Most chemicals that will kill bacteria, viruses, and parasites will also kill fish, or will contaminate them and cause a problem when they are consumed. Salt can be effective in treating parasites in freshwater fish and there are a few other chemicals that are approved for use in treating fish disease. The best way is to prevent an infection by keeping the tanks and equipment clean and by being careful about introducing new organisms from other farms into the system. Other means to treat the water are being tried, including ultraviolet light and ozone. These treatments can be effective, but they are very expensive. This type of aquaculture requires tight monitoring and a high level of expertise.

Major Cultured Species

Innumerable aquatic species are farmed in small quantities around the world. Major aquaculture industries around the world include the following:

- Salmonidae: Atlantic salmon (Salmo salar) and Rainbow trout (Oncorhynchus mykiss). Also smaller volumes of a variety of other salmonids. Originally developed in Norway, Denmark, and Scotland, now farmed in significant quantities in Europe, Canada, Chile, and Australia (Tasmania).
- Shrimp: Mostly Black tiger shrimp (Penaeus monodon) and increasingly White shrimp (Litopenaeus vannamei). Techniques originally developed in Japan and Taiwan. Mostly farmed through tropical and sub-tropical Asia and South America.
- Carp: European carp, Chinese carps (Grass, Silver and Black), and Indian major carps. Easily the largest global aquaculture industry by volume of production. Major producers are China, India, Southeast Asia, and Europe. Carps, which are herbivores, are major contributors of high quality protein to the diets of poorer people around the world, but their value in commercial markets is low.
- Seaweeds: Many species. Huge volumes, low economic value. Mostly farmed in Asia; particularly Japan, Korea, and China.
- Catfish: Major species are Vietnamese basa, Channel catfish, and African and Asian walking catfish (Clarias batrachus). Mostly farmed in Asia and the Southern United States.

- Tilapia: Nile tilapia and a few other species. An herbivorous species very well suited to subsistence farming, although arguably not well suited to large aquabusiness due to finicky breeding biology and low flesh recovery (although becoming a very successful import in the United States and Europe). Mostly farmed in Asia, South America, and Africa.
- Oysters: Pacific oyster (Crassostrea gigas), American oyster (Crassostrea virginica), Flat oyster (Ostrea edulis), and others. Mostly farmed in Asia, United States, Australia, New Zealand, and Europe. Flat oyster was once a huge industry and low cost/very high quality food for the masses in Europe, but collapsed under mortalities brought about by the parasite Bonamia.
- Mussels: Blue mussel (Mytilus edulis), Green mussels (Perna sp.) Mostly farmed in Europe, Asia, New Zealand, and South America.
- Tuna: Southern Bluefin tuna, Northern Bluefin tuna. Tuna farming in Australia has had immense financial success. Tuna farming at present is really a fattening enterprise, where wild bred juvenile tuna are captured and grown in pens to a larger size and better flesh quality. Having the fish confined in pens also means that harvests can be timed to suit the market. This practice has resulted (at least in Australia) on reduced pressure on wild populations and a much larger value for their relatively small wild (Southern bluefin) tuna quota.

Challenges

Like other agriculture production, aquaculture must stand up to a rigorous evaluation of any environmental impact. For example, Salmon aquaculture has come under increasing scrutiny from environmental nongovernmental organizations (ENGOs). In Canada, salmon farming sites occupy a small portion of the coastal zone areas where they are located. The total area occupied by Canadian salmon farms in British Columbia and the Bay of Fundy in New Brunswick is less than 0.01 percent of the coastal area where these sites are located. Still, even though salmon farms occupy only a small percentage of the public waters, scientists have found a significant degradation of the areas where they exist, with lowered oxygen levels, replacement of native seaweeds with invasive seaweeds, increased algal blooms, reduction of wild species, and loss of nursery habitat for wild fish.

Many farmed fish species are carnivorous, meaning that other wild fish species must be harvested in order to maintain the fish farm. For example, herring are used to make salmon feed. Since herring are the backbone of the North Atlantic food chain, increased fishing pressure on their numbers is a serious threat to all other fish species, and other species such as seals, that depend on herring for food. It is argued that fish farms, far from removing the pressure on wild fish stocks, increase it. Others argue that it takes less fish (in the form of the fishmeal component of an aquaculture diet) to produce a unit of table fish through aquaculture than through the natural food web. Fisheries that are based on species lower on the trophic web (such as many species used for fishmeal) are also more resistant to overfishing than typical table fish fisheries.

The fish farm industry is trying to decrease its reliance on fish for fish feed. The vast majority of aquaculture production on the global scale involves omnivorous species such as carp, catfish, and tilapia), which can be raised on feeds using very little or no fishmeal. A portion of the fishmeal used in fish feeds for highly carnivorous species comes from the trimmings and discards of commercial species.

More studies are being done concerning shifts in feed composition using poultry and vegetable oils as substitutes for fish protein and oil. However this use of land-based feed ingredients results in a decrease of the Omega 3 fish oils in the farmed fish (although in some cases a 'washing out' of the terrestrial oils can be achieved with a short period of feeding with marine oils prior to harvest). The current reluctance to further reduce fishmeal and marine oils in the commercial diets of species such as the salmonids and shrimps is based not so much on technical difficulties as on consumer resistance to the taste and health qualities of vegetarian fish. In the long term, alternative sources of long-chain Omega 3 fatty acids (the most difficult ingredient to acquire from non-fish sources) may be developed from zooplankton or microalgal origins.

Other problems with aquaculture include the potential for increasing the spread of unwanted invasive species, as farmed species are often not native to the area in which they are being farmed. When these species escape, as tilapia has done in Florida due to flooding, they can compete with native species and damage ecosystems. Another problem is the spread of introduced parasites, pests, and diseases.

While the negative impacts of some aquaculture on the environment have been widely publicized, the positive environmental effects of aquaculture are often overlooked. For example, many aquacultured species are highly sensitive to water quality conditions and aquaculture farmers often notice the effects of pollution or reductions in water quality before other authorities. Aquaculture businesses have a vested interest in clean waterways, in that a reduction in water quality has a direct effect on their production rates and financial profitability. Appropriate aquacultural development can serve as 'canaries' for the health of waterways, with farms often conducting very regular and quite sophisticated monitoring of their aquatic environment.

Purpose of Aquaculture

Aquaculture serves many purposes, including:

- Food production for human consumption;
- Rebuilding of populations of threatened and endangered species;

- Habitat restoration;
- Wild stock enhancement;
- Production of baitfish; and
- Fish culture for zoos and aquariums.

It is one of the fastest growing forms of food production in the world. Because harvest from many wild fisheries has peaked globally, aquaculture is widely recognized as an effective way to meet the seafood demands of a growing population.

Using aquaculture techniques and technologies, researchers and the aquaculture industry are "farming" all types of freshwater and marine species of fish and shellfish:

- Marine aquaculture refers specifically to the culturing of oceanic species (as opposed to freshwater). Examples of marine aquaculture production include oysters, clams, mussels, shrimp, salmon and algae. Marine aquaculture is just 20 percent of U.S. production, consisting mostly of shellfish (e.g., oysters, clams and mussels).
- Freshwater aquaculture includes trout, catfish and tilapia. About 70 percent of aquaculture in the United States is freshwater farming of catfish and trout. Only a handful of U.S. farms grow marine finfish such as salmon in Maine and Washington State and yellowtail and Pacific threadfin (moi) in Hawaii.

Economic Benefits and Importance of Aquaculture

Alternative Food Source

Fish and other seafood are good sources of protein. They also have more nutritional value like the addition of natural oils into the diet such as omega 3 fatty acids. Also since it offers white meat, it is better for the blood in reducing cholesterol levels as opposed to beef's red meat. Fish is also easier to keep compared to other meat producing animals as they are able to convert more feed into protein. Therefore, its overall conversion of pound of food to pound of protein makes it cheaper to rear fish as they use the food more efficiently.

Alternative Fuel Source

Algae are slowly being developed into alternative fuel sources by having them produce fuels that can replace the contemporary fossil fuels. Algae produce lipids that if harvested can be burn as an alternative fuel source whose only by products would be water when burnt.

Such a breakthrough could ease the dependency of the world on drilled fossil fuels as well as reduce the price of energy by having it grown instead of drilling petroleum.

Moreover, algae fuel is cleaner and farmable source of energy, which means it can revolutionize the energy sector and create a more stable economy that avoids the boom-bust nature of oil and replaces it with a more abundant fuel source.

Increase Jobs in the Market

Aquaculture increases the number of possible jobs in the market as it provides both new products for a market and create job opportunities because of the labor required to maintain the pools and harvest the organisms grown. The increase in jobs is mostly realized in third world countries as aquaculture provides both a food source and an extra source of income to supplement those who live in these regions.

Aquaculture also saves fishermen time as they do not have to spend their days at sea fishing. It allows them free time to pursue other economic activities like engaging in alternative businesses. This increase in entrepreneurship provides more hiring possibilities and more jobs.

Reduce Sea Food Trade Deficit

The sea food trade in America is mainly based on trade from Asia and Europe, with most of it being imported. The resultant balance places a trade deficit on the nation. Aquaculture would provide a means for the reduction of this deficit at a lower opportunity cost as local production would mean that the sea food would be fresher. It would also be cheaper due to reduce transport costs.

Environmental Benefits

Creates Barrier against Pollution with Mollusc and Sea Weed

Molluscs are filter feeders while seaweed acts a lot like the grass of the sea. Both these organisms sift the water that flows through them as brought in by the current and clean the water. This provides a buffer region that protects the rest of the sea from pollution from the land, specifically from activities that disturb the sea bed and raise dust.

Also, the economic benefits of molluscs and sea weed can create more pressure from governments to protect their habitats as they serve an economic importance. The financial benefits realised provides incentive for the government to protect the seas in order to protect sea food revenue.

Reduces Fishing Pressure on Wild Stock

The practice of aquaculture allow for alternative sources of food instead of fishing the same species in their natural habitats. Population numbers of some wild stocks of some species are in danger of being depleted due to overfishing.

Aquaculture provides an alternative by allowing farmers to breed those same species in captivity and allow the wild populations to revitalize. The incentive of less labor for more gains pushes fishers to convert to fish farmers and make even more profit that before. It also allows them control of the supply of the fish in the market giving them the ability to create surplus stock or reduce their production to reap the best profits available.

Importance of Aquaculture

Sustainable use of Sea Resources

Aquaculture provides alternatives for fishing from the sea. Increase in demand for food sources and increase in globalization has led to increase in fishing. Yet, this has led fishermen to become selfish and overfish the desired or high-demand species. Through aquaculture, it provides both an alternative and opportunity for wild stocks to replenish overtime.

Conservation of Biodiversity

Aquacultures also protect biodiversity by reducing the fishing activities on wild stock in their ecosystems. By providing alternatives to fishing, there is reduced attack on the wild populations of the various species in the sea. Reduced action of fishing saves the diversity of the aquatic ecosystem from extinction due to overfishing.

Increased Efficiency, More Resources for Less Effort

Fish convert feed into body protein more efficiently than cattle or chicken production. It is much more efficient meaning that the fish companies make more food for less feed. Such an efficiency means that less food and energy is used to produce food, meaning that the production process is cheaper as well. It saves resources and even allows for more food to be produced leading to secure reserves and less stress on the environment.

Reduced Environmental Disturbance

By increasing aquaculture, fish farming in specific, there is a reduced need for the fishing of the wild stock. As an outcome, it puts less stress on the ecosystem and equally reduces human interference. Actions of motor boats and other human influences such as the removal of viable breeding adult fish are all stresses put on the aquatic ecosystems and their discontinuation allows the ecosystem to flourish and find their natural balance.

Branches of Aquaculture 2

- Fish Farming
- Shrimp Farming
- Oyster Farming
- Algaculture
- Organic Aquaculture
- Offshore Aquaculture
- Coral Aquaculture
- Inland Saline Aquaculture
- Integrated Multi-trophic Aquaculture
- Mariculture

There are a numerous branches of aquaculture such as fish farming, shrimp farming, oyster farming, mariculture, offshore aquaculture, coral aquaculture, inland saline aquaculture and algaculture. The topics elaborated in this chapter will help in gaining a better perspective about these branches of aquaculture.

Fish Farming

Fish farming is a form of aquaculture in which fish are raised in enclosures to be sold as food. It is the fastest growing area of animal food production. Today, about half the fish consumed globally are raised in these artificial environments. Commonly farmed species include salmon, tuna, cod, trout and halibut. These "aquafarms" can take the form of mesh cages submerged in natural bodies of water, or concrete enclosures on land.

According to the United Nations Food and Agriculture Organization, roughly 32% of world fish stocks are overexploited, depleted or recovering and need of being urgently rebuilt. Fish farming is hailed by some as a solution to the overfishing problem. However, these farms are far from benign and can severely damage ecosystems by introducing diseases, pollutants and invasive species. The damage caused by fish farms varies, depending on the type of fish, how it is raised and fed, the size of the production, and where the farm is located.

One significant issue is that—rather than easing the impact on wild populations—the farms often depend on wild fish species lower on the food chain, like anchovies, in order to feed the larger, carnivorous farmed species. It can take up to five pounds of smaller fish to produce one pound of a fish like salmon or sea bass. Overfishing of these smaller "forage" fish has repercussions throughout the ocean ecosystem.

Shrimp Farming

Marine shrimp farming is a century-old practice in many Asian countries. Until a decade ago, this commodity was generally considered a secondary crop in traditional fish farming practices. Shrimp fry trapped in salt beds, coastal paddy fields or brackishwater fishponds are allowed to grow to marketable size and harvested as secondary crop. However, in recent years when higher income are derived from the harvest of shrimp than the principal crop, many farmers have converted their rice fields, salt beds and fishponds into shrimp farms.

In the traditional farming system, the ponds are stocked with fry either collected from the wild or concentrated through tidal water entering the ponds. Shrimp production is inconsistent and varies from year to year due to the dependence on seasonal supply of fry from the wild. Pond yield is also low (100–300 kg/ha/ year) because of inefficient control of predators and competitors, full dependence on natural food and inadequate pond depth.

Some improvements of the traditional farming methods have been made in the past years. Stocking density of shrimp ponds can be increased through concentration of fry by pumping more tidal water into the pond. Pond depth is increased to minimize fluctuations of environmental parameters. As a result, pond yield has correspondingly increased. However, expansion of the shrimp farming industry is still restricted due to the inconsistency in fry supply.

The success in the mass production of hatchery-bred shrimp fry in the 1970's has accelerated shrimp farming development in the region. With improved pond culture techniques, yield from traditional shrimp ponds has been raised to 500–800 kg/ha/ year without supplementary feeding. Pond yield can be further increased to 5–10 tons through supplementary feeding and intensive pond management.

The long gestation period in the development of shrimp farming practice may be due to inadequate technical and financial inputs to effectively demonstrate its commercial viability. Shrimp farming has now developed into an important export-oriented food industry especially in South Asian countries. The perception of an unlimited market demand, high export price, generation of employment and increase in foreign exchange earnings may have encouraged many countries in the region rich in aquatic resources to place high emphasis on the development of the shrimp culture industry.

Practices for Shrimp Farming

External Nursery

External Nurseries allow shrimp farmers to closely monitor baby shrimp. Before external nurseries, farmers would put all the juvenile shrimp into the pond directly, and if an illness hit the shrimp, farmers would quickly lose their crop. By using a nursery system, farmers are able to closely monitor their juvenile shrimp, helping to reduce illness and enabling farmers to keep a close eye on their crop. Shrimp are most vulnerable in their first 40 days of life, and by keeping juveniles out of large ponds, farmers are able to prevent the spread of germs that occur naturally in the surrounding environment. This best practice improves the survival rate of shrimp, the sustainability of farms and the income of farmers. This is most critical practice for raising shrimp, and farmers who incorporate external nurseries are far more successful and exponentially reduce their risk.

Central Drain Systems

In order to prevent illness, farmers have realized the importance of keeping a clean and healthy farming environment. Central drain systems clear sediment and detritus material from the bottom of the pond. Farm workers operate the central drain multiple times per day to remove this waste, helping to create a clean environment. This technology keeps the pond water clean, improves survivability and enhances the sustainability of a shrimp farm. Farms with a central drain are more successful and more sustainable.

Crab and Bird Nets

Crab and bird nets are protective nets that keep predatory species out of shrimp farms. Rogue crabs from local waterways can destroy pond infrastructure, carry illnesses, and force the emergency harvests of shrimp before they are of optimal size. Additionally, local aquatic birds can eat shrimp. Increasing the biosecurity of shrimp farms improves the overall health of the shrimp, reduces the chances that they fall victim to predators, and mitigates the destruction of pond infrastructure. Farms with this infrastructure are better able to raise shrimp and have higher survivability rates.

Aerators

Aerators typically consist of spinning paddle wheels or oxygen lines that increase the dissolved oxygen in a pond. This equipment is pivotal to shrimp farming as it enables us to grow more shrimp, more efficiently, in higher densities. Ponds with high levels of dissolved oxygen decrease the overall stress of a shrimp, lowering the risk of illnesses and increasing growth rates. Additionally, ponds that use the best practice of central drains use circulating paddle wheels to consolidate debris into the central drain for removal. Aerators encourage the growth of healthy shrimp.

Pond Liners

Pond liners reduce erosion on a shrimp farm, especially by aerator systems that circulate the water. Fully lined ponds can also reduce water seepage and the amount of water needed to raise shrimp. There is, however, a debate among farmers about whether partially lined or fully lined ponds are more sustainable – fully lined ponds reduce the amount of seepage into the earth, but partially lined ponds allow shrimp to eat algae and small insects that thrive on bottom of a pond (reducing the amount of feed necessary to grow shrimp). All agree that the use of pond liners help reduce edge erosion, decrease maintenance costs and improve the longevity of the ponds.

Oyster Farming

The most traditional form of husbandry, or aquaculture, is the practice of moving wild oysters as small juveniles from public bottom to leased areas for grow-out. This practice has gone on for centuries and is still practiced today in many areas such as the Gulf of Mexico and the Chesapeake Bay. In addition, clean oyster shells or other media are planted in areas to attract natural recruitment. However, pollution and overfishing have modified the way oysters are produced to meet the marketplace demand. The Chesapeake Bay once was one of the most prolific oyster grounds in the U.S., but oyster disease, unregulated harvest, and the decline in water quality led to the decimation of the oyster populations. Now with better management in place, including an effort to restore public bottom with suitable substrate, harvesting wild oysters in the Chesapeake Bay is still viable, although not at the level it once was.

Need for Hatcheries

Meeting the marketplace demand now requires the use of land-based hatchery technology to produce oyster seed in addition to natural recruitment in the wild. This demand has created a new form of oyster aquaculture utilizing hatchery-produced seed.

In some places, such as the Pacific Northwest, almost all the commercially grown oysters are from hatcheries. The hatcheries spawn the oysters and grow the larvae in tanks until they are ready to set. There are two main oyster culture methods to choose from, either single set or remote setting. Single set culture uses a small cultch sized to encourage development of single oysters for the half-shell market; most of this type of oyster seed is grown out in bags, racks, or rafts. Remote setting typically uses large tanks filled with oyster shells as cultch. This cultch can be sprayed onto large open-water leases and the oysters grown out on the bottom.

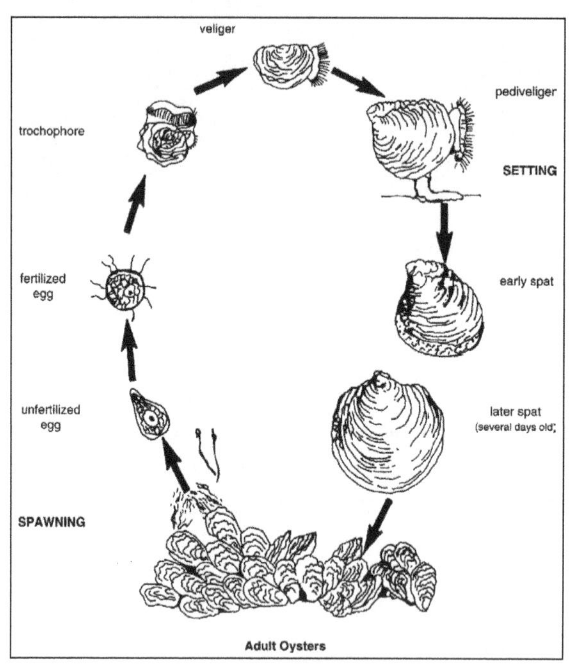

Oyster Life Cycle

Broodstock

To produce broods of larvae in the hatchery, sexually mature adult oysters, known as broodstock, must be available. Broodstock can be obtained from the wild or reared as offspring from previous years. A major factor in selecting offspring is genetics. Scientists throughout agriculture have been able to create disease-resistant or faster-growing organisms through the process of selective breeding. Thus, it is very important to consider the genetics of the broodstock that you are using and to know their origin.

After broodstock are selected, the next step is to determine ripeness, or gonadal maturity. Unfortunately, the only way to tell if a group of oysters is ripe is to shuck a few and examine the gonadal tissue to determine if they are ready to be spawned (i.e., if they possess mature gametes) or if they need to be conditioned. Conditioning is the process of manipulating environmental conditions in the hatchery, such as water temperature, to mimic the environmental conditions needed for gonad maturity. This is a common hatchery technique used to extend the production of oysters outside of their natural spawning season. The basic conditioning techniques involve placing broodstock oysters in tanks containing water at ambient temperature and gradually increasing the water temperature and algal supply over several weeks to ripen the oysters. The water temperature and conditions will vary based on the region and location of the oysters.

Spawning Oysters

Although some oyster culturists still rely on the natural spawn in the wild, the oysters produced from hatcheries are produced through either induced spawning or strip spawning.

Induced Spawning

An induced spawn requires that you have ripe oysters that are ready to spawn. Induced spawning is the process of manipulating water temperature to simulate the oysters' natural spawning conditions. This process is typically done in a spawning table, which is a large, shallow tank set up to circulate warmed seawater from a sump back into the table. This process can take minutes or hours. Sacrificing a male oyster to suspend the sperm in the water can help trigger females to spawn more quickly. Once spawning begins, individual oysters are moved to separate containers of warm seawater to collect the eggs and sperm to keep the eggs from being fertilized until the desired moment. Once the oysters stop spawning, the eggs are collected by pouring the water from containers over a fine mesh screen and are resuspended in clean seawater, when sperm are added to fertilize the eggs. After early development is determined microscopically, the newly formed embryos are placed in tanks at a density of 10/mL.

Oyster spawning

The advantage of induced spawning is the availability to use the same broodstock in subsequent spawns.

A potential disadvantage is that timing is not under your control; the time it takes for the spawn to initiate and the number of broodstock participating can vary greatly.

Strip Spawning

Strip spawning involves removing eggs or sperm from shucked, ripe oysters using a scalpel. A small sample of gonadal tissue is smeared on a glass slide and microscopically examined to determine the gender of each oyster. These oysters are then selected for the quality of their sperm or eggs. The stripped gonadal tissue is washed with seawater over a 75-micron screen to remove debris, while the eggs and sperm are collected separately and then mixed similar to the induced spawning method.

The advantage of a strip spawn is the timing; as long as the oysters are ripe, the spawn takes only as long as it takes to sacrifice the oysters. The disadvantage is the broodstock are sacrificed and not available for the future.

Once you have spawned your oysters, the next step is raising them to the desired market size. Newly "hatched" oysters are born with a shell; larval oysters will be about 20 micrometers in diameter with a filamentous velum that allows the larvae to swim. Raising baby oysters, also known as larviculture, requires providing them with proper food and water quality. Food supply is one of the most critical components to any hatchery, and for oysters the food supply is microscopic algae.

Algal Culture

The foundation for any hatchery operation is the availability of food. For oysters, this food is microalgae and lots of it. There are two main methods used to culture algae in an oyster hatchery: batch culture and continuous culture.

The batch culture method is the process of culturing algae in increasingly larger volumes. Starting from flask stock cultures and moving up to large tanks, each smaller culture is moved into a larger vessel or tank to achieve the volume of algae required for hatchery operations.

The continuous culture method begins the same as the batch culture method, starting with flask stock cultures and then eventually moving to a larger vessel. Instead of moving cultures to larger vessels or tanks, however, water is continuously added to keep the algae culture in the exponential growth phase. Algae are also harvested continuously so that there is an equal amount of water going in and coming out, thus keeping the culture at a constant volume.

Whether algae are produced through continuous or batch culture, the food supply is a vital part of any hatchery and is the key to successfully conditioning oysters for spawning and larval rearing.

Oyster Larviculture

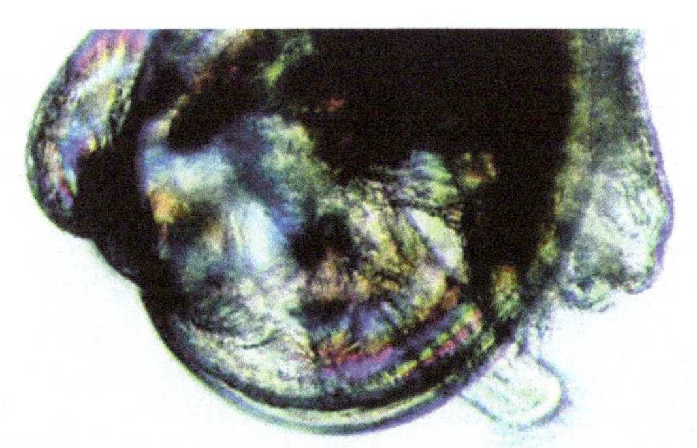

Pictomicrograph of eyes larvae with pseudopod extended

Larviculture, or the raising of baby oysters, is a process that takes about two weeks to 20 days depending on water temperature. Larval oysters are commonly reared in aerated, static cultures from which the larvae are harvested every other day to check survival and growth. The tanks are cleaned and restocked with larvae using progressively lower densities, with algal rations added daily. Flow-through larval-rearing systems are more challenging to manage and require more algal production but allow for larvae to be cultured at much higher densities (<100/ml), dramatically increasing larval production. The larvae will be cared for in this manner depending on how long the larvae take to develop an eye spot on their shells and a foot, also known as "eyed" larvae or "pediveligers," when they are considered competent or ready to set.

Table: Larviculture parameters.

Days after spawn	Larval height (μm)	Sieve size (μm) (diagonal)	Larval density (larvae)/mL	Algal density (cells/mL)
0	–	20	10	20-25,000
2	65	35(50)	5	20-25,000
4	100	53(75)	5	20-25,000
6	140	53(75)	4-5	30-40,000
8	180	73(103)	4	50,000
10	220	73(103)	4	50,000
12	260	100(141)	3	70-80,000
14	290	100(141)	2.5	100-150,000

Setting

Once the larvae have a foot, they will begin looking for a suitable substrate for setting. Once they have set on a substrate, known as cultch, the larvae are then considered to be spat or juvenile oysters and will then be moved to either a nursery system or a grow-out system. There are two main methods for setting oysters, depending on how the oysters will be cultured: remote setting and single spat setting. If remote setting is used to develop spat-on-shell oysters, the eyed larvae will be harvested and transported to be set on cultch material. If intensive culture is the desired grow-out method, the oysters will be set in a way to produce single seed traditionally destined for the half-shell or box market.

Initial set on oyster shell: The small black dots are set oyster spat

Oyster spat after a few weeks in a grow-out system

Setting spat-on-shell can be done at the hatchery or at a remote location by the end user or grower. Containers filled with cleaned oyster shells are placed into large tanks. The types of containers vary from mesh bags to large cages, depending on the volume of production and the level of mechanization. The tanks are filled with filtered seawater, and the eyed larvae are then poured into the tank and allowed to set on the oyster shells. Typically, the system is left closed with only aeration for one to two days to allow the larvae to set. The water can be heated in colder climates or early in the season to allow for a successful set. Then, the system is opened to allow ambient seawater to flow through. The system remains this way for a week or so to allow for spat to grow to a size barely visible to the naked eye. After this time, the tank is drained, and the spatted shells are removed and placed directly on the bottom in open water for grow-out.

Producing single seed oysters for the half-shell market typically takes place in the hatchery, but the larvae can be sold to the end users provided they have the proper equipment. This process requires using micro-cultch, which is finely crushed oyster shells or coral. The cultch is placed on shallow fine mesh screens in shallow tanks of ambient, preferably filtered, seawater. A natural biofilm is allowed to coat the cultch for a day or two prior to adding the larvae. Small volumes of larvae are added and allowed to set on the cultch. The larvae soon outgrow the small fragment to become single oysters. In this method, the small oysters are kept in the hatchery for several weeks and fed a diet of cultured algae before they are released to the nursery system.

Nursery

Many hatcheries care for their juvenile oysters until they are larger before putting them into a grow-out system. This is to ensure better survival and therefore higher production rates. Several types of nursery systems are used, but the concept is the same. All nursery systems pump ambient seawater to provide a constant flow of water and food to the oysters. The designs and locations may differ — some are floating, like the solar-powered floating upwell system (FLUPSY) shown in the adjacent picture, while some are land-based — but the goal is to allow the oysters to grow approximately an inch (from hinge to bill) as quickly as possible. From the nursery, these seed oysters can either be sold to other commercial producers for grow-out or placed in the company's own grow-out facilities.

Solar-powered floating upweller system (FLUPSY)

Grow-out

The harvest and movement of oysters to provide food for human consumption date back centuries, to the time of the Romans. The traditional method for culturing oysters was to simply harvest oysters from the bottom using a pair of tongs. The Industrial Revolution led to the use of faster ships that allowed boats to reach oyster reefs more quickly and to the mechanized winch that allowed for dredging reefs, which was much more efficient than a person using tongs. Today, the traditional method of harvesting seed oysters and moving them to private leases for grow-out is still common in the Gulf of Mexico and the Chesapeake Bay. However, in other regions, hatchery-produced oysters now provide the majority of seed oysters for grow-out.

Two common forms of oyster aquaculture use hatchery products: intensive culture (off bottom, containerized) and extensive culture (loose on bottom). Whether obtaining oyster seed as spat-on-shell or as individuals from a nursery system, the ultimate goal is to grow them to a desirable market size. The market size varies from state to state.

Intensive Culture/Cultchless Production

Intensive culture is also referred to as cultchless production because the oyster seed used is single seed (or without cultch). Cultchless production is containerized for predator protection. Containerization varies from floats, bags, rafts, trays, longline systems, racks, and off-bottom cages. Each of these systems has to face factors that can hinder production, such as impacts from storms, fouling organisms, and conflicts with other users of the waters. Growers select the system based on the water quality, wave action, and predation levels at their grow-out location. Intensive culture is more expensive because it requires more labor for gear and product maintenance. However, the end result is a single, more uniform product selling at a higher price in the boxed or half-shell markets.

Extensive Culture/Cultched Production

Adjustable longline system being researched in Louisiana

Broodstock held in Taylor floats

Rack systems

Oyster condos in Washington

The second aquaculture production method, extensive culture, is referred to as spat-on-shell or cultched production. This method is more traditional in that oyster larvae from a hatchery are set on clean oyster shells (cultch), planted directly on the bottom, and grown out in clusters. This production method is considered relatively less expensive because once planted, there is little to no maintenance. However, mortality is higher overall for extensive culture, and the product is destined for the shucked market, which generally garners a lower price.

Triploid Oysters

Oyster aquaculture is similar to traditional agriculture in that seed is selected for improved survival and performance. Two forms of improvements have been used extensively in the oyster industry: disease resistance and polyploidy. Research has developed disease-resistant strains of oysters through the process of selective breeding. Oyster growth and meat quality have been improved through the use of polyploidy, or the development of a triploid oyster. Triploid (3n) oysters have three sets of chromosomes rather than two sets like a wild organism, which are diploid (2n). Triploid organisms are sterile, which means they do not put energy into reproduction. The energy not used for reproduction results in faster growth rates. Because they are non-reproductive, triploids are marketable in the summertime when natural oysters are not, which greatly expands market potential. Many hatcheries provide a seed that is either disease resistant or triploid or both.

The process of developing triploid oysters was begun through the use of chemicals during the fertilization process. The chemical is used to suppress the first polar body in the cell division of fertilized eggs, resulting in a percentage of triploid offspring. However, oysters produced using chemicals are not approved for human consumption. The next step was to use the chemical triploids to produce tetraploid (4n) oysters (organisms with four sets of chromosomes), which are not sterile and can be cross-bred with diploid oysters to make triploids. Fertilization of diploid oyster eggs with sperm from a tetraploid male results in ~100% triploid offspring, with no use of chemicals. The development of the tetraploid technology revolutionized the oyster industry, and many oyster hatcheries now offer triploid oysters.

Diploid oysters (left) and triploid oysters (right) showing differences in summertime marketability

Algaculture

Algaculture is the farming of algae. High oil prices, competing demands between foods and other biofuel sources, and the world food crisis, have ignited interest in algaculture (farming algae) for making vegetable oil, biodiesel, bioethanol, biogasoline, biomethanol, biobutanol and other biofuels, using land that is not suitable for agriculture. Algae holds enormous potential to provide a non-food, high-yield, non-arable land use source of biodiesel, ethanol and hydrogen fuels. Microalgae are the fastest growing photosynthesizing organism capable of completing an entire growing cycle every few days. Up to 50% of algae's weight is comprised of oil, compared with, for example, oil palm which yields just about 20% of its weight in oil.

Algaculture (farming of algae) can be a route to making vegetable oils, biodiesel, bioethanol and other biofuels. Microalgae are one-celled, photosynthetic microorganisms that are abundant in fresh water, brackish water, and marine environments everywhere on earth. The potential for commercial algae production is expected to come from growth in translucent tubes or containers called photo bioreactors or open ocean algae bloom harvesting. The other advantages of algal systems include:

- Carbon capture from smokestacks to increase algae growth rates.
- Processing of algae biomass through gasification to create syngas.
- Growing carbohydrate rich algae strains for cellulosic ethanol.
- Using waste streams from municipalities as water sources.

Algae have certain qualities that make the organism an attractive option for biodiesel production. Unlike corn-based biodiesel which competes with food crops for land resources, algae-based production methods, such as algae ponds or photobioreactors, would "complement, rather than compete" with other biomass-based fuels. Unlike corn or other biodiesel crops, algae do not require significant inputs of carbon intensive fertilizers. Some algae species can even grow in waters that contain a large amount of salt, which means that algae-based fuel production need not place a large burden on freshwater supplies.

Several companies and government agencies are funding efforts to reduce capital and operating costs and make algae fuel production commercially viable. Companies such as Sapphire Energy and Bio Solar Cellsare using genetic engineering to make algae fuel production more efficient. According to Klein Lankhorst of Bio Solar Cells, genetic engineering could vastly improve algae fuel efficiency as algae can be modified to only build short carbon chains instead of long chains of carbohydrates.

Sapphire Energy also uses chemically induced mutations to produce algae suitable for use as a crop. Some commercial interests into large-scale algal-cultivation systems are looking to tie in to existing infrastructures, such as cement factories, coal power plants, or sewage treatment facilities. This approach changes wastes into resources to provide the raw materials, CO_2 and nutrients, for the system.

Organic Aquaculture

Organic aquaculture is a holistic method for farming marine species in line with organic principles. The ideals of this practice established sustainable marine environments with consideration for naturally occurring ecosystems, use of pesticides, and the treatment of aquatic life. Managing aquaculture organically has become more popular since consumers are concerned about the harmful impacts of aquaculture on themselves and the environment.

The availability of certified organic aquaculture products have become more widely available since the mid-1990s. This seafood growing method has become popular in Germany, the United Kingdom and Switzerland, but consumers can be confused or skeptical about the label due to conflicting and misleading standards around the world.

A certified organic product seal on aquaculture products will mean an accredited certifying body has verified that the production methods meet or exceed a country's standard for organic aquaculture production. Organic regulations designed around soil-based systems don't transfer well into aquaculture and tend to conflict with large-scale, intensive (economically viable) practices/goals. There are a number of problems facing organic aquaculture: difficulty of sourcing and certifying organic juveniles (hatchery or sustainable wild stock); 35-40% higher feed cost; more labour-intensive; time and cost of the certification process; a higher risk of diseases, and uncertain benefits. But, there is a definite consumer demand for organic seafood, and organic aquaculture may become a significant management option with continued research.

A number of countries have created their own national standards and certifying bodies for organic aquaculture. While there is not simply one international organic aquaculture standardization process, one of the largest certification organizations is the Global Trust, which delivers assessments and certifications to match the highest quality organic aquaculture standards. The information regarding these standards is available through a personal inquiry.

Many organic aquaculture certifications address a variety of issues including antibiotic and chemical treatments of fish, unrestrained disposal of fish feces into the ocean, fish feeding materials, the habitat of where and how the fish are raised, and proper handling practices including slaughter. Most Organic Aquaculture certifications follow rather strict requirements and standards. These rules may vary between different countries or certification bodies. This leads to confusion when products are imported from other countries, which can result in a backlash from consumers.

Defining acceptable practices is also complicated by the variety of species - freshwater, saltwater, shellfish, finfish, mollusks and aquatic plants. The difficulty of screening pollutants out of an aquatic medium, controlling the food supplies and of keeping track of individual fish may mean that fish and shellfish stocks should not be classified as 'livestock' at all under regulations. This point further exemplifies the need for widespread aquaculture certification standard.

Challenges and Controversy

There is some controversy over licensing restrictions, as some sea food companies propose that wild caught fish should be classified as organic. While wild fish may be free of pesticides and unsustainable rearing practices, the fishing industry may not necessarily be environmentally sustainable.

The variation in standards, as well as the unknown level of actual compliance and the closeness of investigations when certifying are major problems in consistent organic certification. In 2010, new rules were proposed in the European Union to consistently define the organic aquaculture industry. Canada's General Standards Board's (CGSB) proposed updates to their standards were strongly opposed in 2010 because they allowed antibiotic and chemical treatments of fish, up to 30 percent non-organic feed, deadly and uncontrolled impacts on wild species and unrestrained disposal of fish feces into the ocean. These standards would have certified net pen systems as organic. At the other end of the scale, the extremely strict national legislation in Denmark has made it difficult for the existing organic trout industry to develop.

Potential Alternatives to Non-organic Feed and Waste Removal

One major issue in organic aquaculture production is finding practical and sustainable alternatives to non-organic veterinary treatments, feeds, spat and waste disposal. Potential veterinary alternatives include homeopathic treatments and production-cycle limited allopathic or chemical treatments Current requirements usually stipulate a reduction in unsustainable fishmeal, in favor of organic vegetable and fish by-product replacements. A recent study into organic fish feeds for salmon found that while organic feed provide some benefit to the environmental impact of the fishes' life cycles, the loss of fish meals and oils have a significant negative impact. Another study discovered that certain percentages of dietary protein could be safely replaced.

Not only do the fish have to be organically reared, organic fish feeds need to be developed. Research into ways of decreasing the amount on non-sustainable fishmeal in feed is currently focusing on replacement by organic vegetable proteins. Some organic fish feeds becoming available, and the option of integrated multi-species systems (e.g. growing plants using aquaponics, as well as larvae or other fish). For example, locating a shellfish bed next to a finfish farm to dispose of the waste and provide the shellfish with controlled nutrients.

Certifying Bodies that Cover Organic Aquaculture

Certification body	Countries of operation	No. of certified aquaculture farms	Accredited for grower groups	No. of certified groups	Aquaculture commodities within the scheme	Production (tonnes)
Agrior	Israel	2 + 1 fish feed mill	no	NA	Tilapia, carp, red drum, sea bass, sea bream, Ulva and Ulea seaweed	400
Agri Quality Ltd.	New Zealand, Vanuatu, Cook Islands, Malaysia	yes	Example			
Bioland e.V.	Germany, Austria, Belgium, France, Italy, Netherlands, Switzerland	no	Example			
Debio	Norway	3	no	NA	salmon, trout, cod	trout 0.5 salmon 120 cod 600
Instituto Biodinamico	Brazil, Argentina, Bolivia, Mexico, Paraguay, Uruguay		yes			
Istituto per la Certificazione Etica e Ambientale	Italy, Lebanon, Turkey		yes			
National Association Sustainable Agriculture Australia	Australia, Timor-Leste, Indonesia, Malaysia, Nepal, New Zealand, Papua New Guinea, Samoa, Sri Lanka, Solomon Islands		yes			
Organic Agriculture Certification Thailand	Thailand	1 (not under the IFOAM-accredited scheme)	Example	0	nile tilapia and butter fish	8 000 litres (fish sauce)

Table from IFOAM: Organic schemes.

- United Kingdom the Soil Association.
- Hungary Biokontrol Hungaria.
- Naturland (Association for Organic Agriculture).
- Spain: Voluntary standards set by the Advisory Group CRAE do not cover organic aquaculture.
- New Zealand - BioGro.
- Switzerland - Bio Suisse.

- Nordic countries (Sweden, Norway) as well as Japan, Thailand and Australia - KRAV.

United States Organic Aquaculture Certification

In 2005, with the growing need for a certification process specifically designed for marine-based farming methods, the National Organic Standards Board and the National Organics Program created a working group called the Aquatic Animal Task Force in order to seek recommendations for the new certification process. The task force was meant to be broken into two divisions: wild fisheries and aquaculture, but the wild fisheries group never materialized.

In 2006, the Aquaculture Working Group delivered a report with suggestions for the production and handling of aquatic animals and plants. However, with the complexity and diversity of the marine systems, the group requested more time to explore bivalve mollusks (oysters, clams, mussels and scallops) in depth. The National Organic Standards Board approved the aquaculture standards in 2007 and reconsidered the aquatic animal feed and facilities until they synthesized the public commentary in 2008. In 2010, the NOSB approved the recommendations for the bivalve mollusks section.

Currently, the legal status of using the organic label for aquatic species, and the future of developing U.S. Department of Agriculture (USDA) certification standards for organic aquaculture products and aquatic species, are under review. It is anticipated that the first version of the rule for organic aquaculture will be announced in April or May 2016 with need for approval by the Office of Management and Budget. It is expected to see the final rule in play by late summer or fall of 2016 with organic aquaculture products likely available in store in 2017. The certification is said to include the following: shellfish, marine and recirculating system methods of aquaculture, as well as the controversial net-pen method.

The US currently allows the imports of organically-certified seafood from Europe, Canada and other countries around the world.

Production

Organic aquaculture was responsible for an estimated US$46.1 billion internationally. There were 0.4 million hectares of certified organic aquaculture in 2008 compared to 32.2 million hectares dedicated to Organic farming. The 2007 production was still only 0.1% of total aquaculture production.

The market for organic aquaculture shows strong growth in Europe, especially France, Germany and the UK - for example, the market in France grew 220% from 2007 to 2008. There is a preference for organic food, where available. Organic seafood is now sold in discount supermarket chains throughout the EU. The top five producing countries are UK, Ireland, Hungary, Greece and France. 123 of the 225 global certified organic aquaculture farms operate in Europe and were responsible for 50,000 tonnes in 2008 (nearly half global production).

Organic seafood products are a niche market and users currently expect to pay premiums of 30-40%. Organic salmon is the top species and retails at 50%. Market demand is driving Danish rainbow trout farmers to switch to organic farming.

Offshore Aquaculture

Offshore aquaculture, also known as open ocean aquaculture, is an emerging approach to mariculture or marine farming where fish farms are moved some distance offshore. The farms are positioned in deeper and less sheltered waters, where ocean currents are stronger than they are inshore. Existing 'offshore' developments fall mainly into the category of exposed areas rather than fully offshore. As maritime classification society, DNV GL, has stated, development and knowledge-building are needed in several fields for the available deeper water opportunities to be realized.

One of the concerns with inshore aquaculture is that discarded nutrients and feces can settle below the farm on the seafloor and damage the benthic ecosystem. According to its proponents, the wastes from aquaculture that has been moved offshore tend to be swept away from the site and diluted. Moving aquaculture offshore also provides more space where aquaculture production can expand to meet the increasing demands for fish. It avoids many of the conflicts that occur with other marine resource users in the more crowded inshore waters, though there can still be user conflicts offshore. Critics are concerned about issues such as the ongoing consequences of using antibiotics and other drugs and the possibilities of cultured fish escaping and spreading disease among wild fish.

Aquaculture is the most rapidly expanding food industry in the world as a result of declining wild fisheries stocks and profitable business. In 2008, aquaculture provided 45.7% of the fish produced globally for human consumption; increasing at a mean rate of 6.6% a year since 1970.

In 1970, a National Oceanic and Atmospheric Administration (NOAA) grant brought together a group of oceanographers, engineers and marine biologists to explore whether offshore aquaculture, which was then considered a futuristic activity, was feasible. In the United States, the future of offshore aquaculture technology within federal waters has become much talked-about. As many commercial operations show, it is now technically possible to culture finfish, shellfish, and seaweeds using offshore aquaculture technology.

Major challenges for the offshore aquaculture industry involve designing and deploying cages that can withstand storms, dealing with the logistics of working many kilometers from land, and finding species that are sufficiently profitable to cover the costs of rearing fish in exposed offshore areas.

Technology

To withstand the high energy offshore environment, farms must be built to be more robust than those inshore. However, the design of the offshore technology is developing rapidly, aimed at reducing cost and maintenance.

While the ranching systems currently used for tuna use open net cages at the surface of the sea, the offshore technology usually uses submersible cages. These large rigid cages – each one able to hold many thousands of fish – are anchored on the sea floor, but can move up and down the water column. They are attached to buoys on the surface which frequently contain a mechanism for feeding and storage for equipment. Similar technology is being used in waters near the Bahamas, China, the Philippines, Portugal, Puerto Rico, and Spain. By submerging cages or shellfish culture systems, wave effects are minimized and interference with boating and shipping is reduced. Offshore farms can be made more efficient and safer if remote control is used, and technologies such as an 18-tonne buoy that feeds and monitors fish automatically over long periods are being developed.

Existing Offshore Structures

Multi-functional use of offshore waters can lead to more sustainable aquaculture "in areas that can be simultaneously used for other activities such as energy production". Operations for finfish and shellfish are being developed. For example, the Hubb-Sea World Research Institutes' project to convert a retired oil platform 10 nm off the southern California coast to an experimental offshore aquaculture facility. The institute plans to grow mussels and red abalone on the actual platform, as well as white seabass, striped bass, bluefin tuna, California halibut and California yellowtail in floating cages.

Integrated Multi-Trophic Aquaculture

Integrated multi-trophic aquaculture (IMTA), or polyculture, occurs when species which must be fed, such as finfish, are cultured alongside species which can feed on dissolved nutrients, such as seaweeds, or organic wastes, such as suspension feeders and deposit feeders. This sustainable method could solve several problems with offshore aquaculture. The method is being pioneered in Spain, Canada, and elsewhere.

Roaming Cages

Roaming cages have been envisioned as the "next generation technology" for offshore aquaculture. These are large mobile cages powered by thrusters and able to take advantage of ocean currents. One idea is that juvenile tuna, starting out in mobile cages in Mexico, could reach Japan after a few months, matured and ready for the market. However, implementing such ideas will have regulatory and legal implications.

Space Conflicts

As oceans industrialise, conflicts are increasing among the users of marine space. This competition for marine space is developing in a context where natural resources can be seen as publicly owned. There can be conflict with the tourism industry, recreational fishers, wild harvest fisheries and the siting of marine renewable energy installations. The problems can be aggravated by the remoteness of many marine areas, and difficulties with monitoring and enforcement. On the other hand, remote sites can be chosen that avoid conflicts with other users, and allow large scale operations with resulting economies of scale. Offshore systems can provide alternatives for countries with few suitable inshore sites, like Spain.

Ecological Impacts

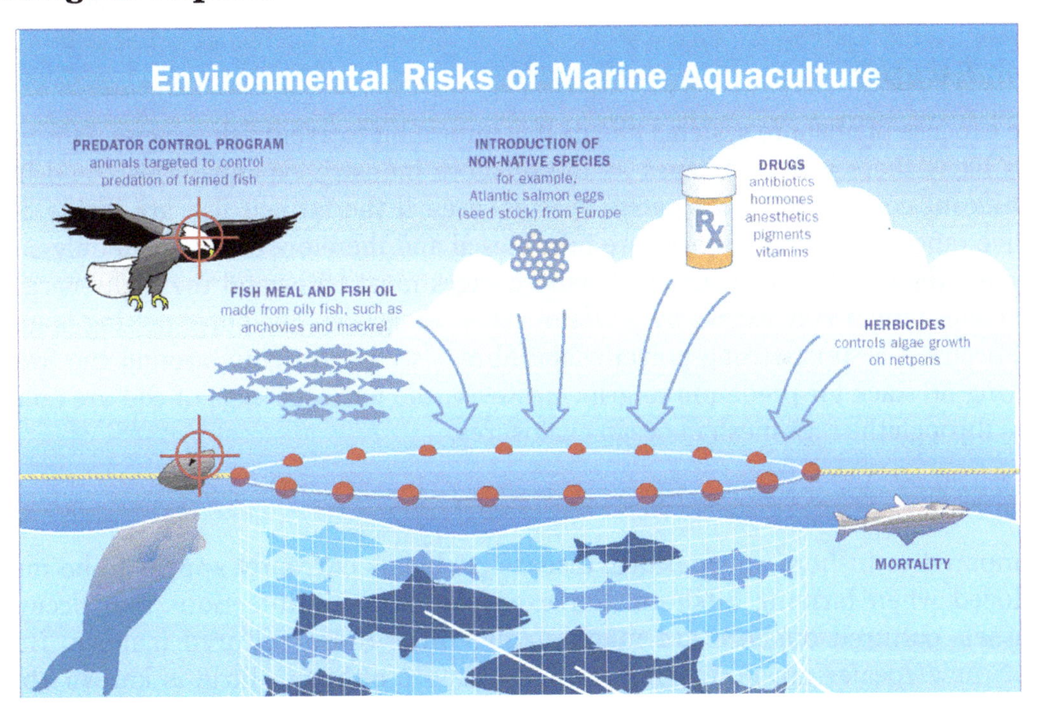

Inshore marine farming systems in shallow sheltered water, as depicted here, can have problems with waste collecting on the sea floor. These problems are lessened with offshore aquaculture, where wastes are swept away from the site and diluted.

The ecological impacts of offshore aquaculture are somewhat uncertain because it is still largely in the research stage. Many of the concerns over potential offshore aquaculture impacts are paralleled by similar, well established concerns over inshore aquaculture practices.

Pollution

One of the concerns with inshore farms is that discarded nutrients and feces can settle on the seafloor and disturb the benthos. The "dilution of nutrients" that occurs in deeper water is a strong reason to move coastal aquaculture offshore into the open ocean. How much nutrient pollution and damage to the seafloor occurs depends on the feed conversion efficiency of the species, the flushing rate and the size of the operation. However, dissolved and particulate nutrients are still released to the environment. Future offshore farms will probably be much larger than inshore farms today, and will therefore generate more waste. The point at which the capacity of offshore ecosystems to assimilate waste from offshore aquaculture operations will be exceeded is yet to be defined.

Wild Caught Feed

As with the inshore aquaculture of carnivorous fish, a large proportion of the feed comes from wild forage fish. Except for a few countries, offshore aquaculture has focused predominantly on high value carnivorous fish. If the industry attempts to expand with this focus then the supply of these wild fish will become ecologically unsustainable.

Fish Escapes

The expense of offshore systems means it is important to avoid fish escapes. However, it is likely there will be escapes as the offshore industry expands. This could have significant consequences for native species, even if the farmed fish are inside their native range. Submersible cages are fully closed and therefore escapes can only occur through damage to the structure. Offshore cages must withstand the high energy of the environment and attacks by predators such as sharks. The outer netting is made of Spectra – a super-strong polyethylene fibre – wrapped tightly around the frame, leaving no slack for predators to grip. However, the fertilised eggs of cod are able to pass through the cage mesh in ocean enclosures.

Disease

Compared to inshore aquaculture, disease problems currently appear to be much reduced when farming offshore. For example, parasitic infections that occur in mussels cultured offshore are much smaller than those cultured inshore. However, new species are now being farmed offshore although little is known about their ecology and epidemiology. The implications of transmitting pathogens between such farmed species and wild species "remains a large and unanswered question".

Spreading of pathogens between fish stocks is a major issue in disease control. Static offshore cages may help minimize direct spreading, as there may be greater distances between aquaculture production areas. However, development of roaming cage technology could bring about new issues with disease transfer and spread. The high level of carnivorous aquaculture production results in an increased demand for live aquatic animals for production and breeding purposes such as bait, broodstock and milt. This can result in spread of disease across species barriers.

Employment

Aquaculture is encouraged by many governments as a way to generate jobs and income, particularly when wild fisheries have been run down. However, this may not apply to offshore aquaculture. Offshore aquaculture entails high equipment and supply costs, and therefore will be under severe pressure to lower labor costs through automated production technologies. Employment is likely to expand more at processing facilities than grow-out industries as offshore aquaculture develops.

Prospects

Norway and the United States are currently making the main investments in the design of offshore cages.

FAO

In 2010, the Food and Agriculture Organization (FAO) sub-committee on aquaculture made the following assessments:

Most Members thought it inevitable that aquaculture will move further offshore if the world is to meet its growing demand for seafood and urged the development of appropriate technologies for its expansion and assistance to developing countries in accessing them. Some Members noted that aquaculture may also develop offshore in large inland water bodies and discussion should extend to inland waters as well. Some Members suggested caution regarding potential negative impacts when developing offshore aquaculture.

The sub-committee recommended the FAO "should work towards clarifying the technical and legal terminology related to offshore aquaculture in order to avoid confusion."

Europe

The European Commission issued the following policy statement on aquaculture:

> "Fish cages should be moved further from the coast, and more research and development of offshore cage technology must be promoted to this end. Experience from outside the aquaculture sector, e.g. with oil platforms, may well feed into the aquaculture equipment sector, allowing for savings in the development costs of technologies."

European offshore systems were operating in Norway, Ireland, Italy, Spain, Greece, Cyprus, Malta, Croatia, Portugal and Libya.

In Ireland, as part of their National Development Plan, it is envisioned that over the period 2007–2013, technology associated with offshore aquaculture systems will be developed, including: "sensor systems for feeding, biomass and health monitoring, feed control, telemetry and communications and cage design, materials, structural testing and modelling."

United States

Moving aquaculture offshore into the exclusive economic zone (EEZ) can cause complications with regulations. In the United States, regulatory control of the coastal states generally extends to 3 nm, while federal waters (or EEZ) extend to 200 nm offshore. Therefore, offshore aquaculture can be sited outside the reach of state law but within federal jurisdiction. As of 2010, "all commercial aquaculture facilities have been sited in nearshore waters under state or territorial jurisdiction." However, "unclear regulatory processes" and "technical uncertainties related to working in offshore areas" have hindered progress. The five offshore research projects and commercial operations in the US – in New Hampshire, Puerto Rico, Hawaii and California – are all in federal waters. In June 2011, the National Sustainable Offshore Aquaculture Act of 2011 was introduced to the House of Representatives "to establish a regulatory system and research program for sustainable offshore aquaculture in the United States exclusive economic zone".

Current Species

By 2005, offshore aquaculture was present in 25 countries, both as experimental and commercial farms. Market demand means that the most offshore farming efforts are directed towards raising finfish. Two commercial operations in the US, and a third in the Bahamas are using submersible cages to raise high-value carnivorous finfish, such as moi, cobia, and mutton snapper. Submersible cages are also being used in experimental systems for halibut, haddock, cod, and summer flounder in New Hampshire waters, and for amberjack, red drum, snapper, pompano, and cobia in the Gulf of Mexico.

The offshore aquaculture of shellfish grown in suspended culture systems, like scallops and mussels, is gaining ground. Suspended culture systems include methods where the shellfish are grown on a tethered rope or suspended from a floating raft in net containers. Mussels in particular can survive the high physical stress levels which occur in the volatile environments that occur in offshore waters. Finfish species must be feed regularly, but shellfish do not, which can reduce costs. The University of New Hampshire in the US has conducted research on the farming of blue mussels submerged in an open ocean environment. They have found that when farmed in less polluted waters offshore, the mussels develop more flesh with lighter shells.

Coral Aquaculture

Coral aquaculture, also known as coral farming or coral gardening, is the cultivation of corals for commercial purposes or coral reef restoration. Aquaculture is showing promise as a tool for restoring coral reefs, which are dying off around the world. The process protects young corals while they are most at risk of dying. Small corals are propagated in nurseries then replanted on the reef.

Coral farmers live near the reefs they farm and work for reef conservation or for income. Coral is also farmed by scientists for research, by businesses for the live and ornamental coral trade, and by private aquarium hobbyists.

Coral reef farming is the extracting of part of a coral colony or free-floating larva from a reef and growing them in a nursery until adulthood. It is commonly referred to as the "gardening method" and has been compared to silviculture as a management practice that mimics natural ecosystems.

Adult corals can be transplanted onto a reef, usually a damaged area. Coral is farmed for conservation reasons in the Philippines, Solomon Islands, Palau, Fiji, Marshall Islands and Japan. Land-based coral farming occurs in public aquariums in North America and Europe.

Reefs in Decline

Coral is an important foundation species. While it covers less than one percent of the ocean surface, it provides habitat for nearly one third of saltwater fish species, as well as ten percent of all fish captured for human consumption.

Coral in an aquaculture tank

Reefs are affected by severe weather events, such as cyclones, from predation by crown of thorns starfish and from competition for habitat with other foundation species such as seaweed. Seaweed can take over coral habitat when the water contains excess nutrients (nitrogen and phosphorus) or when fishing stocks are too low and herbivorous fish do not keep the seaweed at bay by eating it.

Natural stressors to the coral reef are further aggravated by the human impact on coral reefs. Anthropogenic stressors such as runoff, coastal development, dynamite fishing, cyanide fishing, overexploitation of resources and marine pollution, put 58% percent of the world's reefs under threat as of 2009. An example is the exploitation of mushroom coral in Indonesia which is harvested for supply of the jewelry and curio trades. Harvesting of living reef organisms, including coral, is increasing around the world. Coral is often overharvested to supply growing demand. Overharvesting weakens the ability of reefs to replenish after other harmful events.

Reef Restoration

Coral aquaculture/transplantation can improve coral cover, biodiversity, and structural heterogeneity of a degraded reef. Success has been achieved with fire coral, Pocillopora verrucosa and Acropora hemprichii. A restored reef hosts organisms associated with the reef, such as reef fishes.

Nursery-grown coral promote reef resilience by making contributions to the larval pool. This could have a positive effect on new growth if transplanting of the new coral is made just before a larval release season.

Oceanographer Baruch Rinkevich coined the term active restoration to describe coral reef farming, in contrast with what he described as passive restoration efforts focused on mitigation of stressors by means such as the designation of marine protected areas (MPAs). Coral reefs are often placed in MPAs in the hope that reducing human activity will allow the coral to recover.

Aquarium Trade

Coral in a culture facility

Many people enjoy creating their own coral display in a home aquarium. In response to this, businesses farm coral to supply them. Some companies farm in sunlit greenhouses instead of artificially lighted aquariums. The 1999 Hawaii Marine Ornamentals Conference concluded with a recommendation to "give highest priority to projects involving the advancement of marine ornamental aquaculture and reef preservation." Conferees pressed the importance of encouraging hobbyists to supply only from coral reef farms to help deter over-harvesting. Conferees recommended initiatives to encourage consumer understanding that cultured ornamentals are a more sustainable and 'higher value' alternative to wild-caught live reef organisms.

Methods

Coral fragments recovered from bomb fishing sites ready for replanting

Coral fragments replanted in nontoxic cement

Conservation

The stages to farming for reef restoration are: collecting polyps or larvae; growing the specimens in tanks; further growth in sea nurseries and re-transplantation onto the reef.

Collection

Coral can reproduce asexually by budding or sexually by spawning. Collecting coral polyps from existing reef colonies or fragments can be done any time. Branches, fragments, or tips of branches are common targets. This is the most widely practiced method.

Collecting coral spawn is generally an annual activity, conducted immediately following a spawning event. Coral colonies on a reef usually spawn together in a synchronized event on a specific day. This allows for hundreds of thousands of coral embryos to be collected at one time. This method is known as *spat stocking*.

At the Great Barrier Reef Aquarium in Townsville, Australia, large colonies of *Acropora formosa* have collection devices placed above them during spawning. Small mature colonies are transplanted from the reef into a tank for spawning. They can then be reattached to the reef.

Using this method, the mother colonies are unaffected. This method has also been proved effective on Red Sea soft coral species, *Alcyonarians*: *Clavularia hamra*, *Nephthea sp.* and *Litophyton arboreum*.

Tank Cultivation

Linden describes an apparatus made of Petri dishes lined with preconditioned Mailer's paper disks on which the planula of *Stylophora pistillata* are grown. One-month-old survivors were transferred onto plastic pins in a mid-water coral nursery, where the trays were covered with fitted plastic nets to prevent predation and detachment. After four months, more than 89% of the corals had survived.

Ocean Cultivation

Next the corals are transported into floating nurseries in the sea. The corals float in the water column, attached on a submerged structure. Some authors recommend 6 metres depth to ensure the corals get the right amount of sunlight. They are affixed to an artificial substrate. This is usually made from string, wire, mesh, monofilament line or epoxy. The colonies remain there from 8 to 24 months to reach a size for transplantation back to the reef.

Return to the Reef

When the corals are big enough to be transplanted into the reef, the transplantation stage involves securing to the corals by plastic pegs or masonry anchors or with epoxy.

For Commercial or Exhibition Supply

For commercial markets, the process is the same except that the ocean cultivation is extended until the colonies reach marketable size (about fist-sized) and the final step is replaced by extraction and packaging for sale.

Economy

Coral aquaculture offers alternative livelihoods to people living near the reefs. This is especially important for communities where fishing or harvesting marine organisms have become unsustainable, such as in Indonesia. It is possible to use coral resources in a way that is environmentally friendly. Many coral reefs are in impoverished locations. Coral reef aquaculture requires only basic, cheap materials, making it possible for communities with limited resources. Some new methods, such as seeding of tetrapods containing coral larvae, make it possible to reduce costs and outplanting time compared to previous approaches.

Research and Development

Coral aquaculture provides insights into coral life histories. Petersen showed that early sexual recruits grow larger when fed the nauplii of brine shrimp. This discovery could shorten the fragile post settlement time in the hatchery.

The Mote Marine Laboratory keeps many broodstock colonies at its Tropical Research Laboratory. The laboratory website reports that its colonies are grown from fragments rescued from boat groundings and environmental disturbances. The corals in the broodstock reserve provide fragments for restoration research. Studies are done to determine optimal size, shape and season for restoration.

Market

Indonesia and the Philippines supply ~85% of coral reef products. Indonesia requires 10% of coral production to be transplanted into the ocean. As of 2012, a majority of coral imports to the US were wild-caught, although an increasing proportion were cultured. From 1990 to 2010, imports increased by some 8% annually. Imports declined thereafter the wake of the Great Recession and from increasing domestic production. Commercial trade in stony and reef-building corals is regulated by the Convention on International Trade in Endangered Species (CITES). In Indonesia, most production is located around airports to speed the shipping process.

Inland Saline Aquaculture

Inland saline aquaculture is the farming or culture of aquatic animals and plants using inland (i.e. non-coastal) sources of saline groundwater rather than the more common coastal aquaculture methods. As a side benefit, it can be used to reduce the amount of salt in underground water tables, leading to an improvement in the surrounding land usage for agriculture. Due to its nature, it is only commercially possible in areas that have large reserves of saline groundwater, such as Australia.

Systems

Extensive Culture

Extensive culture aquaculture systems are simple and with low levels of intervention. An example of this would be a salty dam, lake or pond stocked with trout, where no food is needed to be added as the fish can feed off what naturally occurs in the water. While they required little capital investment or management time, their productivity is relatively low.

Intensive Culture

Intensive culture requires more capital outlay and greater management time. Often they use purpose-built facilities (e.g. tanks), artificial food and aeration and constant monitoring of water quality. It has much higher productivity rates, but associated high levels of feeding, labour, water pumping and capital costs.

Semi-Intensive Culture

Semi-intensive culture is in between extensive and intensive culture. It may range from adding some artificial feed to an extensive system or some aeration and waste management. Costs rise as more inputs are added.

Suitable Species

Fish

- Rainbow trout: Robust, fast growth, require low water temperatures, may be limited to winter production.
- Brown trout: Robust, fast growth, require low water temperatures, may be limited to winter production.
- Barramundi: Needs higher temperatures, tolerant in a large range of salinity levels.
- Macquarie perch: Wide tolerance over range of salinity and water quality levels, not suitable for commercial quantities.
- Silver Perch: Suitable for extensive and intensive systems, prefers warmer water.
- Snapper.

Other Species

- Crustaceans: Brine, shrimp, prawns - these can be included as part of a wastewater treatment program as some have the capacity to quickly clean water.
- Molluscs: Mussels.
- Algae: Both unicellular and "Seaweeds" can be used to extract a range of high-value products, including pharmaceutical chemicals.

Mixing Species

Chain System

Some inland aquaculture systems involve using a range of separated species to increase its productivity. An example of this would be where water is used to culture a fish specifies, which is then diverted to tanks of shellfish which feed on the fine particles left by the fish, which then is diverted to algae species which remove the dissolved nutrients, and then last of all the water is sent to a horticultural system.

Poly-culture

Separate from this type of system is poly-culture, where two or more species are cultured in the same water, possibly multiple fish species or a fish and mollusc species.

Integrated Multi-trophic Aquaculture

Integrated multi-trophic aquaculture (IMTA) provides the byproducts, including waste, from one aquatic species as inputs (fertilizers, food) for another. Farmers combine fed aquaculture (e.g., fish, shrimp) with inorganic extractive (e.g., seaweed) and organic extractive (e.g., shellfish) aquaculture to create balanced systems for environment remediation (biomitigation), economic stability (improved output, lower cost, product diversification and risk reduction) and social acceptability (better management practices).

Selecting appropriate species and sizing the various populations to provide necessary ecosystem functions allows the biological and chemical processes involved to achieve a stable balance, mutually benefiting the organisms and improving ecosystem health.

Ideally, the co-cultured species each yield valuable commercial "crops". IMTA can synergistically increase total output, even if some of the crops yield less than they would, short-term, in a monoculture.

"Integrated" refers to intensive and synergistic cultivation, using water-borne nutrient and energy transfer. "Multi-trophic" means that the various species occupy different trophic levels, i.e., different (but adjacent) links in the food chain.

IMTA is a specialized form of the age-old practice of aquatic polyculture, which was the co-culture of various species, often without regard to trophic level. In this broader case, the organisms may share biological and chemical processes that may be minimally complementary, potentially leading to reduced production of both species due to competition for the same food resource. However, some traditional systems such as polyculture of carps in China employ species that occupy multiple niches within the same pond, or the culture of fish that is integrated with a terrestrial agricultural species, can be considered forms of IMTA.

The more general term "Integrated Aquaculture" is used to describe the integration of monocultures through water transfer between the culture systems. The terms "IMTA" and "integrated aquaculture" differ primarily in their precision and are sometimes interchanged. Aquaponics, fractionated aquaculture, integrated agriculture-aquaculture systems, integrated peri-urban-aquaculture systems, and integrated fisheries-aquaculture systems are all variations of the IMTA concept.

Range of Approaches

Today, low-intensity traditional/incidental multi-trophic aquaculture is much more common than modern IMTA. Most are relatively simple, such as fish, seaweed or shellfish.

True IMTA can be land-based, using ponds or tanks, or even open-water marine or freshwater systems. Implementations have included species combinations such as shellfish/shrimp, fish/seaweed/shellfish, fish/seaweed, fish/shrimp and seaweed/shrimp.

IMTA in open water (offshore cultivation) can be done by the use of buoys with lines on which the seaweed grows. The buoys/lines are placed next to the fishnets or cages in which the fish grows. In some tropical Asian countries some traditional forms of aquaculture of finfish in floating cages, nearby fish and shrimp ponds, and oyster farming integrated with some capture fisheries in estuaries can be considered a form of IMTA. Since 2010, IMTA has been used commercially in Norway, Scotland, and Ireland.

In the future, systems with other components for additional functions, or similar functions but different size brackets of particles, are likely. Multiple regulatory issues remain open.

Modern History of Land-based Systems

Ryther and co-workers created modern, integrated, intensive, land mariculture. They originated, both theoretically and experimentally, the integrated use of extractive organisms—shellfish, microalgae and seaweeds—in the treatment of household effluents, descriptively and with quantitative results. A domestic wastewater effluent, mixed with seawater, was the nutrient source for phytoplankton, which in turn became food for oysters and clams. They cultivated other organisms in a food chain rooted in the farm's organic sludge. Dissolved nutrients in the final effluent were filtered by seaweed (mainly Gracilaria and Ulva) biofilters. The value of the original organisms grown on human waste effluents was minimal.

In 1976, Huguenin proposed adaptations to the treatment of intensive aquaculture effluents in both inland and coastal areas. Tenore followed by integrating with their system of carnivorous fish and the macroalgivore abalone.

In 1977, Hughes-Games described the first practical marine fish/shellfish/phytoplankton culture, followed by Gordin, et al., in 1981. By 1989, a semi-intensive (1 kg fish/m^{-3}) seabream and grey mullet pond system by the Gulf of Aqaba (Eilat) on the Red Sea supported dense diatom populations, excellent for feeding oysters. Hundreds of kilos of fish and oysters cultured here were sold. Researchers also quantified the water quality parameters and nutrient budgets in (5 kg fish m^{-3}) green water seabream ponds. The phytoplankton generally maintained reasonable water quality and converted on average over half the waste nitrogen into algal biomass. Experiments with intensive bivalve cultures yielded high bivalve growth rates. This technology supported a small farm in southern Israel.

Sustainability

IMTA promotes economic and environmental sustainability by converting byproducts and uneaten feed from fed organisms into harvestable crops, thereby reducing eutrophication, and increasing economic diversification.

Properly managed multi-trophic aquaculture accelerates growth without detrimental side-effects. This increases the site's ability to assimilate the cultivated organisms, thereby reducing negative environmental impacts.

IMTA enables farmers to diversify their output by replacing purchased inputs with by-products from lower trophic levels, often without new sites. Initial economic research suggests that IMTA can increase profits and can reduce financial risks due to weather, disease and market fluctuations. Over a dozen studies have investigated the economics of IMTA systems since 1985.

Nutrient Flow

Typically, carnivorous fish or shrimp occupy IMTA's higher trophic levels. They excrete soluble ammonia and phosphorus (orthophosphate). Seaweeds and similar species can extract these inorganic nutrients directly from their environment. Fish and shrimp also release organic nutrients which feed shellfish and deposit feeders.

Species such as shellfish that occupy intermediate trophic levels often play a dual role, both filtering organic bottom-level organisms from the water and generating some ammonia. Waste feed may also provide additional nutrients; either by direct consumption or via decomposition into individual nutrients. In some projects, the waste nutrients are also gathered and reused in the food given to the fish in cultivation. This can happen by processing the seaweed grown into food.

Recovery Efficiency

Nutrient recovery efficiency is a function of technology, harvest schedule, management, spatial configuration, production, species selection, trophic level biomass ratios, natural food availability, particle size, digestibility, season, light, temperature, and water flow. Since these factors significantly vary by site and region, recovery efficiency also varies.

In a hypothetical family-scale fish/microalga/bivalve/seaweed farm, based on pilot scale data, at least 60% of nutrient input reached commercial products, nearly three times more than in modern net pen farms. Expected average annual yields of the system for a hypothetical 1 hectare (2.5 acres) were 35 tonnes (34 long tons; 39 short tons) of seabream, 100 tonnes (98 long tons; 110 short tons) of bivalves and 125 tonnes (123 long tons; 138 short tons) of seaweeds. These results required precise water quality control and attention to suitability for bivalve nutrition, due to the difficulty in maintaining consistent phytoplanton populations.

Seaweeds' nitrogen uptake efficiency ranges from 2-100% in land-based systems. Uptake efficiency in open-water IMTA is unknown.

Food Safety and Quality

Feeding the wastes of one species to another has the potential for contamination, although this has yet to be observed in IMTA systems. Mussels and kelp growing adjacent to Atlantic salmon cages in the Bay of Fundy have been monitored since 2001 for contamination by medicines, heavy metals, arsenic, PCBs and pesticides. Concentrations are consistently either non-detectable or well below regulatory limits established by the Canadian Food Inspection Agency, the United States Food and Drug Administration and European Community Directives. Taste testers indicate that these mussels are free of "fishy" taste and aroma and could not distinguish them from "wild" mussels. The mussels' meat yield is significantly higher, reflecting the increase in nutrient availability. Recent findings suggest mussels grown adjacent to salmon farms are advantageous for winter harvest because they maintain high meat weight and condition index (meat to shell ratio). This finding is of particular interest because the Bay of Fundy, where this research was conducted, produces low condition index mussels during winter months in monoculture situations, and seasonal presence of Paralytic Shellfish Poisoning (PSP) typically restricts mussel harvest to the winter months.

Mariculture

Mariculture is the farming of aquatic plants and animals in salt water. Thus, mariculture represents a subset of the larger field of aquaculture, which involves the farming of both fresh-water and marine organisms. The major categories of mariculture species are seaweeds, mollusks, crustaceans, and finfish.

Recent information indicates that the total amount of seafood (including fresh-water species and aquatic plants) is about 140 million metric tons annually. Over 20 percent of the total comes from aquatic plants (mostly seaweeds). Marine fish account for only 2 percent of the total.

Mollusks (clams, oysters, abalone, scallops, and mussels) represent the most important species cultured in marine waters. Seaweeds (brown, red, and green) are a close second. While most people do not think that they eat much (or any) seaweed, extracts from seaweeds can be found in everything from toothpaste and ice cream to automobile tires. Seaweeds themselves are dried and used directly as human food in many parts of the world.

Crustaceans include shrimp, crabs, lobsters, and crayfish. While shrimp culture has become a major industry in Asia and Latin American since the early 1980s, global production is far less than that of mollusks and seaweeds. Marine fish production is even smaller. Top finfish groups include Atlantic salmon, milkfish, sea bream, sea bass, red drum, yellowtail, striped bass, and hybrid striped bass.

Table: The top mariculture-producing countries include the following.

The top mariculture producing countries include the following.	
Country	Species Produced
China	mollusks, shrimp
Japan	algae, mollusks, yellowtail, sea bream
Taiwan	mollusks, shrimp, eels
Philippines	algae, shrimp, milkfish
United States	mollusks, shrimp atlantic salmon, red drum
Norway	salmon
Ecuador	shrimp
Republic of Korea	algae, mollusks
Indonesia	algae, shrimp, milkfish

Types of Operations

The culture of blue mussels on long ropes is common in the bays and inlets of Nova Scotia, Canada. This mollusk is economically important to local growers, even though it represents only a small fraction of the province's mollusk production

Various levels of technology are involved in mariculture, the lowest giving nature the major role in producing the crop. The culturist may help prepare the growing area but does little else. For example, oyster culturists may place old shells on the bottom to provide places for a new generation of oysters to attach. The oysters feed on wild phytoplankton and are harvested when they reach the proper size. The next level would be to spawn oysters in a hatchery and allow the larval oysters (called spat) to settle on oyster shell, after which the shell is placed on the oyster bed in bays or suspended on ropes from a raft. Mussels and scallops also can be grown on ropes below rafts.

Ponds

Shrimp and various species of marine fishes are often grown in ponds. The young shrimp and fish are usually produced in hatcheries, though collection of young animals from nature has been used in the past and is still used in some cases. The ponds may be filled with sea water by pumping water, or through tidal flow (the farmer opens the floodgate when the tide is rising and closes it when the pond is full). Depending on the particular species being produced and the size at stocking, the time required for the animals to reach market size can range from a few months to nearly 2 years.

Pens and Cages

In addition to ponds, marine fish also are being reared in floating pens or cages in protected bays. Most cultured salmon are produced in these types of facilities, primarily in Norway, Canada, the United States, Scotland, and Chile. Various other fish species also are being produced in pens and cages in Japan, Europe, and the Middle East. In recent years, there has been interest and a limited amount of activity associated with cage culture in offshore waters.

Indoor Facilities

The highest level of technology is associated with indoor facilities in which the animals are grown in raceways or tanks (circular raceways) that receive pumped seawater that may be taken directly from the ocean. The water may be flowed through the tanks and discarded, or it may be recirculated, that is, reused by passing it through an elaborate water treatment system. Marine species can be reared to market size in such facilities, but they are most commonly used as hatcheries and to hold broodstock.

Considerations

While a number of species are being reared successfully by mariculturists, several desirable ones have not yet been produced economically. This lack of commercial production is because their life cycles either are difficult to control under culture conditions or are very complex. In addition, a number of popular food animals are highly cannibalistic. Various species of crabs and lobsters, for example, are difficult-to-rear species that also are cannibalistic.

Low-technology forms of mariculture take advantage of the nutrients and other physical and chemical characteristics of the ocean. This employee works with cultured seaweed on an East Africa plantation. The cycle of planting and harvesting occurs every 3 to 4 weeks year-round, and daily hours are spent preparing, planting, and collecting the seaweed during low tide.

Opposition to mariculture has developed in several countries since the 1980s. Many people do not want to see pens and cages in their bays, and they are concerned about possible environmental impacts associated with mariculture. Scientists are attempting to address these and a variety of other issues that have been raised. The goal is to produce high-quality seafood in an environmentally responsible manner.

Although world fish production from capture fisheries leveled off during the 1990s, demand for seafood continues to increase. This is because of the growth of the human population and also the view that seafood is healthy food. Scientists believe that natural production from the ocean will not increase; consequently, if the demand for seafood by humans is to be met in the future, both mariculture and fresh-water aquaculture production will have to increase significantly.

References

- Fish-farming, content: awionline.org, Retrieved 4 January, 2019
- Oyster-culture: marine-aquaculture.extension.org, Retrieved 17 March, 2019
- "safeguarding the environment: canadian aquaculture industry alliance". Www.aquaculture.ca. Archived from the original on 2016-03-23. Retrieved 2016-04-22
- Algal-biomass: bioenergyconsult.com, Retrieved 31 July, 2019

- Pelletier, n; tyedmers, p (2007), "feeding farmed salmon: is organic better?", aquaculture, 272 (2): 399–416, doi:10.1016/j.aquaculture.2007.06.024
- "safeguarding the environment: canadian aquaculture industry alliance". Www.aquaculture.ca. Archived from the original on 2016-03-23. Retrieved 2016-04-23
- Lunger, Angela N.; Craig, S. R.; McLean, E. (June 2006), "Replacement of fish meal in cobia (Rachycentron canadum) diets using an organically certified protein", Aquaculture, 257 (1–4): 393–399, doi:10.1016/j.aquaculture.2005.11.010, archived from the originalon 2013-02-02
- Garthwaite, josie (december 17, 2012). "reimagining the coral market". New york times. Retrieved 2018-08-27
- Mariculture, la-mi: waterencyclopedia.com, Retrieved 31 July, 2019

Species Farmed in Aquaculture 3

- Aquatic Plants
- Fish
- Crustaceans
- Molluscs
- Jellyfish
- Echinoderm

There are various species which are bred, reared and harvested in aquaculture. The common species dealt with in aquaculture are aquatic plants, fish, crustaceans, molluscs, echinoderm, jellyfish, etc. The topics elaborated in this chapter will help in gaining a better perspective about these species.

Aquatic Plants

Aquatic plants are those that can be found in any salt or freshwater environment – a small fish tank, home aquarium, lake, pond, ocean, you get the idea. Whether they live above water, are fully submerged in water or somewhere in between doesn't make too much of a difference; the main concept to know is that aquatic plants are any species that naturally prosper in a wet environment.

Characteristics of Aquatic Plants

The characteristics of aquatic plants can vary depending on the type, but there are a few that remain the same. For example, aquatic plants can utilize less of their resources for the purpose of support tissues as they are naturally able to stay afloat. Furthermore, water loss doesn't have to be a concern because these plants are constantly surrounded by it. But aside from those characteristics that are consistent, here are some that vary based on classification.

1. Submerged plants are usually without a cuticle layer in order to avoid excessive dryness.
2. Submerged plants lack xylem since their leaves can do all of the work.
3. The leaves of submerged plants rarely have stomata.
4. Emersed plants have leaves that stick out of the water with access to the air and sun, though their roots are always located at the bottom of a body of water.
5. Free-floating plants have leaves that float on the surface of the water as opposed to sticking out of it.

Benefits of Aquatic Plants

Aquatic plants aren't just used to give a fish tank some character or to serve as a resting place for frogs (though their ability to add to the aesthetics of bodies of water is certainly something worth mentioning). Rather, they are an extremely important part of aquatic environments as they can:

1. Provide protection to fish.
2. Increase oxygen levels in water.
3. Filter water.
4. Prevent undesirable plants from growing.
5. Act as food.

Additionally, in larger bodies of water specifically, they can also help our shores fight against aggressive currents and erosion.

Types of Aquatic Plants

Aquatic Plants for a Fish Tank/Aquarium

Regardless of whether you have a small fish tank at home or support a large aquarium, some popular aquatic plants to consider are water hyacinths, anubias and java moss.

1. Water hyacinths, also known as eichhornia crassipes, are native to South America. A free-floating plant, these are known to be approximately three feet tall at their highest. Water hyacinths are particularly important because they are good aerators for the water and can also absorb harmful waste.

2. Anubias plants are a great choice if you're looking to spice up your fish tank's personality. Known to originate in Africa, these plants really stand out among other fish tank decor as they are one of the taller aquatic plants.

3. Java moss is beneficial for a home fish tank because it can thrive under various conditions and can help maintain the nutrients in water. Additionally, it can grow pretty quickly.

Aquatic Plants for the Ocean

If you've been to the beach, you must be familiar with at least one aquatic plant – seaweed! This type of aquatic plant essentially encompasses several types of marine plants and algae, two being phytoplankton and kelp. Here's how they differ:

1. Phytoplankton, sometimes called microalgae, are one of the smaller types of seaweed. And though this is a plant that thrives in the ocean, it is similar to our terrestrial plants since it does require sunlight in order to survive. You can usually find this type of plant towards the ocean's surface, and one of its main functions is to provide food for those animals that live in the sea.

2. Kelp, on the other hand, is a smaller seaweed that can be found in the ocean. More often than not you'll find "kelp forests" in shallow ocean waters. These are particularly important to ocean life because it provides food and shelter.

Aquatic Plants for a Pond

Even if you don't have a pond in your backyard, it's still important to be familiar with this last group of aquatic plants. We bet you would definitely recognize the first – the water lily:

1. Water lilies: These are also called nymphaea odorata, are important to our waters because of the way that they provide shade and help keep things clean (and of course, they are a great place for frogs to kick back and relax). Similar to phytoplankton, water lilies are best suited for shallow areas, and they're also known for the beautiful flower that blooms aside the leaves (just like the lilies you may have in a vase at home).

2. Lotus: This type of aquatic flower is often mistaken for water lilies as they look pretty similar at first glance. Since they do prefer warm weather it can be difficult for this type of plant to thrive in cold temperatures, but when it is blooming it may be one of the most beautiful aquatic plants you see.

3. Arrow arum: These plants are known for the food that they provide for ducks, and again, they primarily thrive in shallow waters.

Aquatic Plant Care

If you are responsible for aquatic plants in any way, shape or form, be it for your home aquarium, outdoor pond, maybe even a garden with aquatic features, it's important to know that caring for this type of plant life may differ from your usual routines. First things first, make sure that you do your research about which plants can survive in the environment you want to put them in – depending on temperatures and lighting, the right plant for you may not be the same as your neighbor's.

Once you've determined which type of aquatic plant or flower you want to use, make sure to check in on them regularly to ensure that everything is going smoothly. For example, you should prune your plants when needed to ensure that your pond or aquarium is nicely manicured, and you must also check the water to make sure it hasn't been overtaken by any gases that can harm your plants. Finally, if your aquatic plants are living outdoors, make the necessary adjustments to bring them indoors when winter arrives, or rearrange their placement in your pond to ensure that winter weather doesn't disrupt them.

Fish

Fish is any of the 34,000 found in the fresh and salt waters of the world. Living species range from the primitive jawless lampreys and hagfishes through the cartilaginous sharks, skates, and rays to the abundant and diverse bony fishes. Most fish species are cold-blooded; however, one species, the opah (Lampris guttatus), is warm-blooded.

Lamprey (Lampetra) on rainbow trout

The term fish is applied to a variety of vertebrates of several evolutionary lines. It describes a life-form rather than a taxonomic group. As members of the phylum Chordata, fish share certain features with other vertebrates. These features are gill slits at some point in the life cycle, a notochord, or skeletal supporting rod, a dorsal hollow nerve cord, and a tail. Living fishes represent some five classes, which are as distinct from one another as are the four classes of familiar air-breathing animals—amphibians, reptiles, birds, and mammals. For example, the jawless fishes (Agnatha) have gills in pouches and lack limb girdles. Extant agnathans are the lampreys and the hagfishes. As the name implies, the skeletons of fishes of the class Chondrichthyes (from chondr, "cartilage," and ichthyes, "fish") are made entirely of cartilage. Modern fish of this class lack a swim bladder, and their scales and teeth are made up of the same placoid material. Sharks, skates, and rays are examples of cartilaginous fishes. The bony fishes are by far the largest class. Examples range from the tiny sea horse to the 450-kg (1,000-pound) blue marlin, from the flattened soles and flounders to the boxy puffers and ocean sunfishes. Unlike the scales of the cartilaginous fishes, those of bony fishes, when present, grow throughout life and are made up of thin overlapping plates of bone. Bony fishes also have an operculum that covers the gill slits.

The study of fishes, the science of ichthyology, is of broad importance. Fishes are of interest to humans for many reasons, the most important being their relationship with and dependence on the environment. A more obvious reason for interest in fishes is their role as a moderate but important part of the world's food supply. This resource, once thought unlimited, is now realized to be finite and in delicate balance with the biological, chemical, and physical factors of the aquatic environment. Overfishing, pollution, and alteration of the environment are the chief enemies of proper fisheries management, both in fresh waters and in the ocean. Another practical reason for studying fishes is their use in disease control. As predators on mosquito larvae, they help curb malaria and other mosquito-borne diseases.

Landing a fish catch in the harbour of Esbjerg, Den

Fishes are valuable laboratory animals in many aspects of medical and biological research. For example, the readiness of many fishes to acclimate to captivity has allowed biologists to study behaviour, physiology, and even ecology under relatively natural conditions. Fishes have been especially important in the study of animal behaviour, where research on fishes has provided a broad base for the understanding of the more flexible behaviour of the higher vertebrates. The zebra fish is used as a model in studies of gene expression.

There are aesthetic and recreational reasons for an interest in fishes. Millions of people keep live fishes in home aquariums for the simple pleasure of observing the beauty and behaviour of animals otherwise unfamiliar to them. Aquarium fishes provide a personal challenge to many aquarists, allowing them to test their ability to keep a small section of the natural environment in their homes. Sportfishing is another way of enjoying the natural environment, also indulged in by millions of people every year. Interest in aquarium fishes and sportfishing supports multimillion-dollar industries throughout the world.

Harlequin fish (Rasbora heteromorpha)

Structural Diversity

Fishes have been in existence for more than 450 million years, during which time they have evolved repeatedly to fit into almost every conceivable type of aquatic habitat. In a sense, land vertebrates are simply highly modified fishes: when fishes colonized the land habitat, they became tetrapod (four-legged) land vertebrates. The popular conception of a fish as a slippery, streamlined aquatic animal that possesses fins and breathes by gills applies to many fishes, but far more fishes deviate from that conception than conform to it. For example, the body is elongate in many forms and greatly shortened in others; the body is flattened in some (principally in bottom-dwelling fishes) and laterally compressed in many others; the fins may be elaborately extended, forming intricate shapes, or they may be reduced or even lost; and the positions of the mouth, eyes, nostrils, and gill openings vary widely. Air breathers have appeared in several evolutionary lines.

Trumpet fish (Aulostomus maculatus)

Many fishes are cryptically coloured and shaped, closely matching their respective environments; others are among the most brilliantly coloured of all organisms, with a wide range of hues, often of striking intensity, on a single individual. The brilliance of pigments may be enhanced by the surface structure of the fish, so that it almost seems to glow. A number of unrelated fishes have actual light-producing organs. Many fishes are able to alter their coloration—some for the purpose of camouflage, others for the enhancement of behavioral signals.

Fishes range in adult length from less than 10 mm (0.4 inch) to more than 20 metres (60 feet) and in weight from about 1.5 grams (less than 0.06 ounce) to many thousands of kilograms. Some live in shallow thermal springs at temperatures slightly above 42 °C (100 °F), others in cold Arctic seas a few degrees below 0 °C (32 °F) or in cold deep waters more than 4,000 metres (13,100 feet) beneath the ocean surface. The structural and, especially, the physiological adaptations for life at such extremes are relatively poorly known and provide the scientifically curious with great incentive for study.

Distribution and Abundance

Almost all natural bodies of water bear fish life, the exceptions being very hot thermal ponds and extremely salt-alkaline lakes, such as the Dead Sea in Asia and the Great Salt Lake in North America. The present distribution of fishes is a result of the geological history and development of Earth as well as the ability of fishes to undergo evolutionary change and to adapt to the available habitats. Fishes may be seen to be distributed according to habitat and according to geographical area. Major habitat differences are marine and freshwater. For the most part, the fishes in a marine habitat differ from those in a freshwater habitat, even in adjacent areas, but some, such as the salmon, migrate from one to the other. The freshwater habitats may be seen to be of many kinds. Fishes found in mountain torrents, Arctic lakes, tropical lakes, temperate streams, and tropical rivers will all differ from each other, both in obvious gross structure and in physiological attributes. Even in closely adjacent habitats where, for example, a tropical mountain torrent enters a lowland stream, the fish fauna will differ. The marine habitats can be divided into deep ocean floors (benthic), mid-water oceanic (bathypelagic), surface oceanic (pelagic), rocky coast, sandy coast, muddy shores, bays, estuaries, and others. Also, for example, rocky coastal shores in tropical and temperate regions will have different fish faunas, even when such habitats occur along the same coastline.

Although much is known about the present geographical distribution of fishes, far less is known about how that distribution came about. Many parts of the fish fauna of the fresh waters of North America and Eurasia are related and undoubtedly have a common origin. The faunas of Africa and South America are related, extremely old, and probably an expression of the drifting apart of the two continents. The fauna of southern Asia is related to that of Central Asia, and some of it appears to have entered Africa. The extremely large shore-fish faunas of the Indian and tropical Pacific oceans comprise a related complex, but the tropical shore fauna of the Atlantic, although containing Indo-Pacific components, is relatively limited and probably younger. The Arctic and Antarctic marine faunas are quite different from each other. The shore fauna of the North Pacific is quite distinct, and that of the North Atlantic more limited and probably younger. Pelagic oceanic fishes, especially those in deep waters, are similar the world over, showing little geographical isolation in terms of family groups. The deep oceanic habitat is very much the same throughout the world, but species differences do exist, showing geographical areas determined by oceanic currents and water masses.

Distinguishing Taxonomic Features

In forming hypotheses about the evolution of fishes and in establishing classifications based on these hypotheses, ichthyologists place special emphasis on the comparative study of the skeleton. There are two primary advantages of this approach. First, direct comparison between extant and fossil groups is possible, the latter usually represented only by bony remains. The second advantage is that the bones of living fishes are relatively easy to observe and to study, compared with other body structures. Proper preservation and special preparation of the nervous system, for example, are difficult and expensive when the fishes being compared are from the far ends of the Earth. In the study of the relationships of species within a group, major use has been made of similarities and differences in the dimensions of external features, such as head and body length, and of counts of external characters, such as teeth, fin rays, and scales. Colour pattern is also important. In recent years, valuable data on classification of fishes have been obtained from studies of comparative behaviour, physiology, genetics and functional anatomy.

Crustaceans

Crustacean is any member of the subphylum Crustacea (phylum Arthropoda), a group of invertebrate animals consisting of some 45,000 species distributed worldwide. Crabs, lobsters, shrimps, and wood lice are among the best-known crustaceans, but the group also includes an enormous variety of other forms without popular names. Crustaceans are generally aquatic and differ from other arthropods in having two pairs of appendages (antennules and antennae) in front of the mouth and paired appendages near the mouth that function as jaws. Because there are many exceptions to the basic features, however, a satisfactory inclusive definition of all the Crustacea is extraordinarily hard to frame.

Size Range and Diversity of Structure

The largest crustaceans belong to the Decapoda, a large order (about 10,000 species) that includes the American lobster, which can reach a weight of 20 kilograms (44 pounds), and the giant Japanese spider crab, which has legs that can span up to 3.7 metres (12 feet). At the other end of the scale, some of the water fleas (class Branchiopoda), such as Alonella, reach lengths of less than 0.25 millimetre (0.009 inch), and many members of the subclass Copepoda are less than one millimetre in length. The range of structure is reflected in the complex classification of the group. Some of the parasitic forms are so modified and specialized as adults that they can only be recognized as crustaceans by features of their life histories.

Distribution and Abundance

Crustaceans are found mainly in water. Different species are found in freshwater, seawater, and even inland brines, which may have several times the salt concentration of seawater. Various species have occupied almost every conceivable niche within the aquatic environment. An enormous abundance of free-swimming (planktonic) species occupies the open waters of lakes and oceans. Other species live at the bottom of the sea, where they may crawl over the sediment or burrow into it. Different species are found in rocky, sandy, and muddy areas. Some species are so small that they live in the spaces between sand grains. Others tunnel in the fronds of seaweeds or into man-made wooden structures. Some members of the orders Isopoda and Amphipoda extend down to the greatest depths in the sea and have been found in oceanic trenches at depths of up to 10,000 metres. Crustaceans colonize lakes and rivers throughout the world, even high mountain lakes at altitudes of 5,000 metres. They range widely in latitude as well: in the high Arctic some crustaceans use the short summer to develop quickly through a generation, leaving dormant stages to overwinter.

A number of crabs are amphibious, being capable of leaving the water to scavenge on land. Some, like the ghost crabs (Ocypode), can run at great speed across tropical beaches. One of the mangrove crabs, Aratus, can climb trees. Some crabs spend so much time away from the water that they are known as land crabs; however, these crustaceans must return to the water when their larvae are ready to hatch. The most terrestrial of the Crustacea are the wood lice (order Isopoda, family Oniscoidea); most live in damp places, although a few isopod species can survive in deserts. In addition to these well-adapted groups, occasional representatives of other groups have become at least semiterrestrial. Amphipods, members of the subclasses Copepoda and Ostracoda, and the order Anomopoda have been found among damp leaves on forest floors, particularly in the tropics.

Importance to Humans

The crustaceans of most obvious importance to humans are the larger species, chiefly decapods. Fisheries in many parts of the world capture shrimps, prawns, spiny lobsters, and the king crab (Paralithodes) of the northern Pacific and its southern counterpart, the centolla, found off the coast of Chile. Many species of true crabs—such as the blue crab, Dungeness crab, and the stone crab, all in North America, and the edible crab of Europe—are valuable sources of food. The most highly prized decapod is probably the true lobster (Homarus species), although overfishing since the early 20th century has greatly diminished the catches of both the North American and the European species. Freshwater crustaceans include crayfish and some river prawns and river crabs. Many species have only local market value. It is probable that no crustaceans are poisonous unless they have been feeding on the leaves or fruits of poisonous plants.

Another crustacean, the large acorn shell (Balanus psittacus), a barnacle (order Cirripedia) measuring up to 27 centimetres (11 inches) in length, is regarded as a delicacy in South America, and a stalked barnacle (Mitella pollicipes) is eaten in parts of France and Spain. In Japan, barnacles are allowed to settle and grow on bamboo stakes, later to be scraped off and crushed for use as fertilizer.

Copepods and krill are important components of most marine food webs. Planktonic (i.e., drifting) copepods, such as Calanus, and members of the order Euphausiacea (euphausiids), or krill, may be present in such great numbers that they discolour large areas of the open sea, thus indicating to fishermen where shoals of herring and mackerel are likely to be found.

The water flea (Daphnia magna) and the brine shrimp (Artemia salina) are used as fish food in aquariums and fish ponds, and the larvae of the latter are widely used as food for the larvae of larger crustaceans reared in captivity. Ostracods, of which numerous fossil and subfossil species are known, are important to geologists and oil prospectors.

Much damage may be done to rice paddies by burrowing crabs of various species and by the mud-eating, shrimplike Thalassina of Malaya. By undermining paddy embankments, they allow water to drain away, thus exposing the roots of the plants to the sun; if near the coast, salt water may thus be allowed to seep into the paddies. Tadpole shrimps (Triops) are often numerous in rice fields, where they stir up the fine silt in search of food, killing many of the plants. Land crabs and crayfish may damage tomato and cotton crops.

Form and Function of External Features

Although crustaceans exhibit a great variety of forms, the basic crustacean body consists of a number of segments, or somites. These somites sometimes are fused to form rigid areas and sometimes are free, linked to each other by flexible areas that allow some movement. Each somite has the potential for bearing a pair of appendages, although in various crustacean groups appendages are missing from certain somites. The appendages are also jointed with flexible articulations.

At the front, or anterior end, of the body there is an unsegmented, presegmental region called the acron. In most crustaceans at least four somites fuse with the acron to form the head. At the posterior end of the body there is another unsegmented region, the telson, that may bear two processes, or rami, which together form the furca. These two processes at the tail end of the body vary greatly in form; in many crustaceans they are short, but in some they may be as long as the rest of the body. The Crustacea as a whole shows great variation in the number of somites and the amount of fusion that has taken place. In the class Malacostraca, which includes the decapods, there is a consistent body plan: the trunk (which follows the head) is divided into two distinct regions, an anterior thorax of eight somites and a posterior abdomen of seven somites, although as a rule only six are evident in the adult. The reproductive ducts of male malacostracans typically open on the last thoracic somite, and the female reproductive ducts open on the sixth thoracic segment.

Blind cave-dwelling Squat Lobster (Munidopsis polymorpha), Lanzarote, Canary Islands, Spain

The carapace is a widespread crustacean feature, arising during development as a fold from the last somite at the back of the head. It may form a broad fold extending toward the rear over the back, or dorsal surface, of the trunk, as in the notostracan tadpole shrimps, but it often encloses the entire trunk, including limbs and gills. In the clam shrimps (orders Spinicaudata and Laevicaudata) and the ostracods, the carapace is split into two "valves," giving the animals a clamlike appearance. In many decapods the carapace projects forward to form a rostrum, which is often sharply pointed and toothed. The carapace is absent from the anostracans, amphipods, isopods, and members of the superorder Syncarida. Barnacles attach permanently to hard surfaces and use their highly modified carapace to form a mantle. The mantle secretes the barnacle's characteristic calcium carbonate shell plates.

Appendages

There is great diversity among crustacean appendages, but it is thought that all the different types have been derived either from the multibranched (multiramous) limb of the class Cephalocarida or from the double-branched (biramous) limb of the class Remipedia. A biramous limb typically has a basal part, or protopodite, bearing two branches, an inner endopodite and an outer exopodite. The protopodite can vary greatly in its development and may have additional lobes on both its inner and outer margin, called, respectively, endites and exites. The walking legs of many malacostracans have become uniramous by failing to develop the exopodite.

Variations in appendage sequence and morphology largely define different crustacean groups. If one starts at the head of a crustacean and works toward the rear, the following appendages are generally encountered: antennae 1, or antennules; antennae 2, or antennae proper; mandibles; maxillae 1, or maxillulae; maxillae 2, or maxillae proper; and a variable number of trunk limbs. The trunk limbs all may be similar, as in the anostracans and the classes Cephalocarida and Remipedia, or they may be differentiated into distinct groups. In the copepods the first pair of trunk limbs is used for food collection. These limbs are called maxillipeds. In the decapods there are three sets of paired maxillipeds. In the copepods the maxillipeds are followed by four pairs of swimming legs; a fifth pair is sometimes highly modified for reproductive purposes and is sometimes reduced to a mere vestige. Behind the decapod maxillipeds there are five pairs of thoracic limbs, a variable number of which may bear pincers, or chelae. In crabs there is a single obvious pair of chelae, but in some of the prawns there may be up to three pairs of less conspicuous pincers. The decapod abdomen normally bears six pairs of biramous appendages, which are used in swimming in many shrimps and prawns, while in the crabs and crayfish the first two pairs in the male are modified to help in sperm transfer during mating. The last pair of abdominal limbs is frequently different from the others and is called the uropods. In shrimps and lobsters the uropods together with the telson form a tail fan.

The appendages change both their form and their function during the life cycles of most crustaceans. In most adults the antennules and antennae are sensory organs, but in the nauplius larva the antennae often are used for both swimming and feeding. Processes at the base of the antennae can help the mandibles push food into the mouth. The mandibles of a nauplius have two branches with a chewing or compressing lobe at the base; they also may be used for swimming. In the adult the mandible loses one of the branches, sometimes retaining the other as a palp, and the base can develop into a powerful jaw. An alternative development is found in some of the blood-sucking parasites, in which the mandibles form needlelike stylets for piercing their hosts.

Exoskeleton

The outer covering of crustaceans is variously called the integument, cuticle, or exoskeleton. It protects the body and provides attachment sites for muscles. The thickness of the cuticle can vary from a thin, flexible membrane, as in some parasitic copepods, to a massive rigid shell, as in crabs. The cuticle is secreted by a single layer of cells called the epidermis. The outermost layer, or epicuticle, lacks the chitin present in the thicker innermost layers, or procuticle. The procuticle is made up of layers of chitin fibres intermeshed with proteins and, in many species, with calcium salts.

A typical crustacean grows in a series of stages, or molts. The hard exoskeleton prevents any increase in size except immediately after molting. The sequence of events during molting can be divided into four main stages: (1) Proecdysis, or premolt, is the period during which calcium is resorbed from the old exoskeleton into the blood. The epidermis separates from the old exoskeleton, new setae form, and a new exoskeleton is secreted. (2) Ecdysis, or the actual shedding of the old exoskeleton, takes place when the old exoskeleton splits along preformed lines. In the lobster it splits between the carapace and the abdomen, and the body is withdrawn through the hole, leaving the old exoskeleton almost intact. In isopods the exoskeleton is cast in two parts; the front portion may be cast several days after the hind part. Immediately after ecdysis the crustacean swells from a rapid intake of water. (3) Metecdysis, or postmolt, is the stage in which the soft cuticle gradually hardens and becomes calcified. At the end of this stage the cuticle is complete. (4) Intermolt is a period of variable duration, from a few days in small forms to a year or more in some of the large forms. Some crustaceans, after passing through a series of molts, reach a stage where they do not molt again; this is called a terminal anecdysis. The molting process is under hormonal control.

Distinguishing Taxonomic Features

In classifying the Crustacea, a variety of characters are important: the form and extent of the carapace, if present; the number of trunk somites, or segments, and how many fuse with the head or with the telson; the number and degree of specialization of the trunk limbs; the presence or absence of paired eyes and of a caudal furca—i.e., a forked-tail process; and the position and kind of respiratory organs. The position of the genital openings, the mode of attachment of the eggs to the female, and the stage at which the first larva hatches may also be significant. Parasitic and sedentary forms may differ markedly as adults from free-living species.

Molluscs

Mollusk, also spelled mollusc is any soft-bodied invertebrate of the phylum Mollusca, usually wholly or partly enclosed in a calcium carbonate shell secreted by a soft mantle covering the body. Along with the insects and vertebrates, it is one of the most diverse groups in the animal kingdom, with nearly 100,000 (possibly as many as 150,000) described species. Each group includes an ecologically and structurally immense variety of forms: the shell-less Caudofoveata; the narrow-footed gliders (Solenogastres); the serially valved chitons (Placophora or Polyplacophora); the cap-shaped neopilinids (Monoplacophora); the limpets, snails, and slugs (Gastropoda); the clams, mussels, scallops, oysters, shipworms, and cockles (Bivalvia); the tubiform to barrel-shaped tusk shells (Scaphopoda); and the nautiluses, cuttlefishes, squids, and octopuses (Cephalopoda).

Size Range and Diversity of Structure

Typical molluscan features have been substantially altered, or even lost, in many subgroups. Among the cephalopods the giant squids (Architeuthis), the largest living invertebrates, attain a body length of eight metres (more than 26 feet); with the tentacle arms extended, the total length reaches to 22 metres. Other cephalopods exceed a length of one metre. Many of the remaining molluscan classes show a large variation in size: among bivalves the giant clam (Tridacna) ranges up to 135 centimetres (four feet) and the pen shell (Pinna) from 40 to 80 centimetres; among gastropods the sea hares (Aplysia) grow from 40 to 100 centimetres and the Australian trumpet, or baler (Syrinx), up to 60 centimetres; among placophores the gumshoe, or gumboot chiton (Cryptochiton), achieves a length up to 30 to 43 centimetres; and, among solenogasters, Epimenia reaches a length of 15 to 30 centimetres. Finally, gastropods of the family Entoconchidae, which are parasitic in echinoderm sea cucumbers, may reach a size of almost 1.3 metres. In contrast, there are also minute members, less than one millimetre (0.04 inch) in size, among the solenogasters and gastropods.

Distribution and Abundance

The mollusks have adapted to all habitats except air. Although basically marine, bivalves and gastropods include freshwater species. Gastropods have also adapted to land, with thousands of species living a fully terrestrial existence. Found on rocky, sandy, and muddy substrata, mollusks burrow, crawl, become cemented to the surface, or are free-swimming.

Mollusks are found worldwide, but there is a preponderance of some groups in certain areas of the world. The close association of many molluscan groups with their food source—whether by direct dependence on a specific food supply (e.g., plant-eating, or herbivores) or by involvement in food chains—limits their geographic distribution; for example, bivalves of the family Teredinidae (shipworms) are associated with wood. In general, cold-water regions support fewer species.

Importance to Humans

Mollusks are of general importance within food chains and as members of ecosystems. Certain species are of direct or indirect commercial and even medical importance to humans. Many gastropod species, for example, are necessary intermediate hosts for parasitic flatworms (class Trematoda, phylum Platyhelminthes), such as the species that cause schistosomiasis in humans. Most bivalves contribute to the organic turnover in the intertidal (littoral) zones of marine and fresh water because, as filter feeders, they filter up to 40 litres (10 gallons) of water per hour. This filtering activity, however, may also seriously interfere with the various populations of invertebrate larvae (plankton) found suspended and free-swimming in the water. One species, the zebra mussel (Dreissena polymorpha), is regarded as a particularly harmful exotic invader. Carried from Europe in ship ballast water, zebra mussels were taken to the Great Lakes in 1986. To date, they have caused millions of dollars in commercial damage by clogging the water pipes of power plants and cooling systems. They are driving many native freshwater bivalve species to extinction.

Many gastropods, bivalves, and cephalopods are a source of food for many cultures and therefore play an important role in the fishing industries of many countries. Many shell-bearing molluscan species are also used to fabricate ornaments and are harvested for the pearl and mother-of-pearl industries.

External Features

The most obvious external molluscan features are the dorsal epidermis called the mantle (or pallium), the foot, the head (except in bivalves), and the mantle cavity. The mantle in caudofoveates and solenogasters is covered by cuticle that contains scales or minute, spinelike, hard bodies (spicules), or both (aplacophoran level). The chitons (class Polyplacophora) develop a series of eight articulating plates or valves often surrounded by a girdle of cuticle with spicules; in all other mollusks, the mantle secretes an initially homogeneous shell. The mantle and shell are laterally compressed in scaphopods and bivalves; in gastropods and cephalopods the head is free of the mantle and shell. In bivalves a dorsal hinge ligament joins two shell valves, which are further held together by two adductor muscles with attachment points on the inner aspect of each valve.

The molluscan body, which contains all the visceral elements (such as the digestive tract, gonads, and heart), is connected to the mantle by dorsoventral musculature. The head, when present, has tentacles called captacula in scaphopods, labial palps in bivalves, head tentacles in gastropods, and arms in cephalopods. The primitive ciliary gliding surface with forward pedal and sole glands is reduced in caudofoveats and some gastropods, as well as in some bivalves, and it is narrowed to a ridged tract in solenogasters as well as some members of the placophore genus Cryptoplax. The foot forms an anteriorly elongated and slendered burrowing organ in scaphopods, is ax-shaped to vermiform in bivalves, and is modified to a siphon or funnel in cephalopods. Among gastropods of the subclass Opisthobranchia, the foot may be extended laterally to form swimming lobes (parapodia), or even flapping wings (in pteropods, or sea butterflies).

The mantle, or pallial, cavity is found between the mantle rim and the body. The pallial complex is a collection of structures at the roof of the mantle cavity and typically contains at least one pair of lamellate gills (ctenidia), a thick layer of glandular epithelium called mucus tracts or hypobranchial glands, and the outlets for the digestive, excretory, and reproductive systems. A loss of the ctenidia (along with the mucus tracts) is seen in scaphopods, advanced gastropods, septibranch bivalves, and solenogasters.

Jellyfish

Jellyfish and sea jellies are the informal common names given to the medusa-phase of certain gelatinous members of the subphylum Medusozoa, a major part of the phylum Cnidaria. Jellyfish are mainly free-swimming marine animals with umbrella-shaped bells and trailing tentacles, although a few are not mobile, being anchored to the seabed by stalks. The bell can pulsate to provide propulsion and highly efficient locomotion. The tentacles are armed with stinging cells and may be used to capture prey and defend against predators. Jellyfish have a complex life cycle; the medusa is normally the sexual phase, the planula larva can disperse widely and is followed by a sedentary polyp phase.

Jellyfish are found all over the world, from surface waters to the deep sea. Scyphozoans (the "true jellyfish") are exclusively marine, but some hydrozoans with a similar appearance live in freshwater. Large, often colorful, jellyfish are common in coastal zones worldwide. The medusae of most species are fast growing, mature within a few months and die soon after breeding, but the polyp stage, attached to the seabed, may be much more long-lived. Jellyfish have been in existence for at least 500 million years, and possibly 700 million years or more, making them the oldest multi-organ animal group.

Jellyfish are eaten by humans in certain cultures, being considered a delicacy in some Asian countries, where species in the Rhizostomae order are pressed and salted to remove excess water. They are also used in research, where the green fluorescent protein, used by some species to cause bioluminescence, has been adapted as a fluorescent marker for genes inserted into other cells or organisms. The stinging cells used by jellyfish to subdue their prey can also injure humans. Many thousands of swimmers are stung every year, with effects ranging from mild discomfort to serious injury or even death; small box jellyfish are responsible for many of these deaths. When conditions are favourable, jellyfish can form vast swarms. These can be responsible for damage to fishing gear by filling fishing nets, and sometimes clog the cooling systems of power and desalination plants which draw their water from the sea.

Names

The name jellyfish, has traditionally been applied to medusae and all similar animals including the comb jellies (ctenophores, another phylum). The term *jellies* or *sea jellies* is more recent, having been introduced by public aquaria in an effort to avoid use of the word "fish" with its connotations of an animal with a backbone, though shellfish, cuttlefish and starfish are not vertebrates either. In scientific literature, "jelly" and "jellyfish" have been used interchangeably. Many sources refer to only scyphozoans as "true jellyfish".

Anatomy

The main feature of a true jellyfish is the umbrella-shaped bell. This is a hollow structure consisting of a mass of transparent jelly-like matter known as mesoglea, which forms the hydrostatic skeleton of the animal. 95% or more of the mesogloea (the tissue that functions as a hydro-static skeleton) consists of water, but it also contains collagen and other fibrous proteins, as well as wandering amoebocytes which can engulf debris and bacteria. The mesogloea is bordered by the epidermis on the outside and the gastrodermis on the inside. The edge of the bell is often divided into rounded lobes known as lappets, which allow the bell to flex. In the gaps or niches between the lappets are dangling rudimentary sense organs known as rhopalia, and the margin of the bell often bears tentacles.

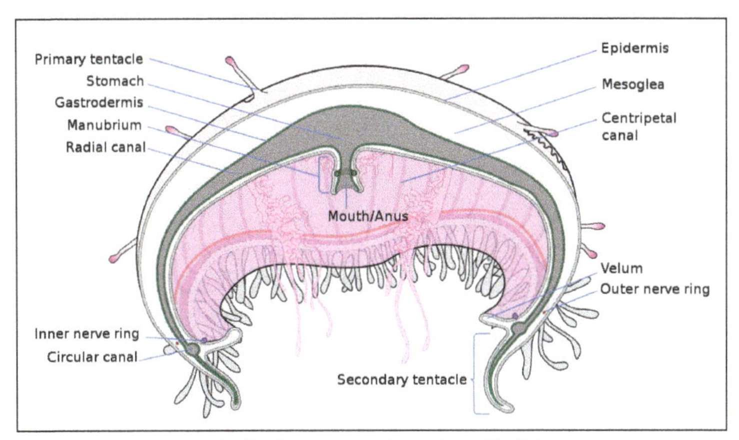

Labelled cross section of a jellyfish

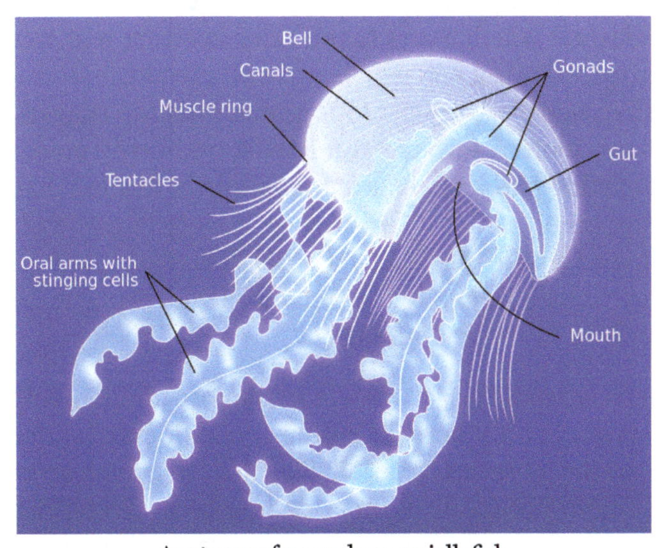

Anatomy of a scyphozoan jellyfish

On the underside of the bell is the manubrium, a stalk-like structure hanging down from the centre, with the mouth, which also functions as the anus, at its tip. There are often four oral arms connected to the manubrium, streaming away into the water below. The mouth opens into the gastrovascular cavity, where digestion takes place and nutrients are absorbed. This is subdivided by four thick septa into a central stomach and four gastric pockets. The four pairs of gonads are attached to the septa, and close to them four septal funnels open to the exterior, perhaps supplying good oxygenation to the gonads. Near the free edges of the septa, gastric filaments extend into the gastric cavity; these are armed with nematocysts and enzyme-producing cells and play a role in subduing and digesting the prey. In some scyphozoans, the gastric cavity is joined to radial canals which branch extensively and may join a marginal ring canal. Cilia in these canals circulate the fluid in a regular direction.

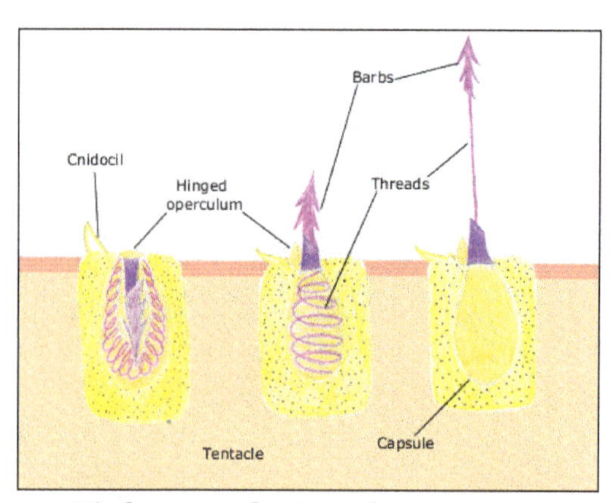

Discharge mechanism of a nematocyst

The box jellyfish is largely similar in structure. It has a squarish, box-like bell. A short pedalium or stalk hangs from each of the four lower corners. One or more long, slender tentacles are attached to each pedalium. The rim of the bell is folded inwards to form a shelf known as a velarium which restricts the bell's aperture and creates a powerful jet when the bell pulsates, allowing box jellyfish to swim faster than true jellyfish. Hydrozoans are also similar, usually with just four tentacles at the edge of the bell, although many hydrozoans are colonial and may not have a free-living medusal stage. In some species, a non-detachable bud known as a gonophore is formed that contains a gonad but is missing many other medusal features such as tentacles and rhopalia. Stalked jellyfish are attached to a solid surface by a basal disk, and resemble a polyp, the oral end of which has partially developed into a medusa with tentacle-bearing lobes and a central manubrium with four-sided mouth.

Most jellyfish do not have specialized systems for osmoregulation, respiration and circulation, and do not have a central nervous system. Nematocysts, which deliver the sting, are located mostly on the tentacles; true jellyfish also have them around the mouth and stomach. Jellyfish do not need a respiratory system because sufficient oxygen diffuses through the epidermis. They have limited control over their movement, but can navigate with the pulsations of the bell-like body; some species are active swimmers most of the time, while others largely drift. The rhopalia contain rudimentary sense organs which are able to detect light, water-borne vibrations, odour and orientation. A loose network of nerves called a "nerve net" is located in the epidermis. Although traditionally thought not to have a central nervous system, nerve net concentration and ganglion-like structures could be considered to constitute one in most species. A jellyfish detects stimuli, and transmits impulses both throughout the nerve net and around a circular nerve ring, to other nerve cells. The rhopalial ganglia contain pacemaker neurones which control swimming rate and direction.

In many species of jellyfish, the rhopalia include ocelli, light-sensitive organs able to tell light from dark. These are generally pigment spot ocelli, which have some of their cells pigmented. The rhopalia are suspended on stalks with heavy crystals at one end, acting like gyroscopes to orient the eyes skyward. Certain jellyfish look upward at the mangrove canopy while making a daily migration from mangrove swamps into the open lagoon, where they feed, and back again. Box jellyfish have more advanced vision than the other groups. Each individual has 24 eyes, two of which are capable of seeing colour, and four parallel information processing areas that act in competition, supposedly making them one of the few kinds of animal to have a 360-degree view of its environment.

Largest and Smallest

Jellyfish range from about one millimeter in bell height and diameter, to nearly 2 metres (6.6 ft) in bell height and diameter; the tentacles and mouth parts usually extend beyond this bell dimension.

The smallest jellyfish are the peculiar creeping jellyfish in the genera *Staurocladia* and *Eleutheria*, which have bell disks from 0.5 millimetres (0.02 in) to a few millimeters in diameter, with short tentacles that extend out beyond this, which these jellyfish use to move across the surface of seaweed or the bottoms of rocky pools. Many of these tiny creeping jellyfish cannot be seen in the field without a hand lens or microscope; they can reproduce asexually by fission (splitting in half). Other very small jellyfish, which have bells about one millimeter, are the hydromedusae of many species that have just been released from their parent polyps; some of these live only a few minutes before shedding their gametes in the plankton and then dying, while others will grow in the plankton for weeks or months. The hydromedusae *Cladonema radiatum* and *Cladonema californicum* are also very small, living for months, yet never growing beyond a few mm in bell height and diameter.

The lion's mane jellyfish (*Cyanea capillata*) is one of the largest species

The lion's mane jellyfish, *Cyanea capillata*, was long-cited as the largest jellyfish, and arguably the longest animal in the world, with fine, thread-like tentacles that may extend up to 36.5 metres (120 ft) long (though most are nowhere near that large). They have a moderately painful, but rarely fatal, sting. The increasingly common giant

Nomura's jellyfish, *Nemopilema nomurai,* found in some, but not all years in the waters of Japan, Korea and China in summer and autumn is another candidate for "largest jellyfish", in terms of diameter and weight, since the largest Nomura's jellyfish in late autumn can reach 2 metres (6 ft 7 in) in bell (body) diameter and about 200 kilograms (440 lb) in weight, with average specimens frequently reaching 0.9 metres (2 ft 11 in) in bell diameter and about 150 kilograms (330 lb) in weight. The large bell mass of the giant Nomura's jellyfish can dwarf a diver and is nearly always much greater than the Lion's Mane, whose bell diameter can reach 1 metre (3 ft 3 in).

The rarely encountered deep-sea jellyfish *Stygiomedusa gigantea* is another candidate for "largest jellyfish", with its thick, massive bell up to 100 centimetres (39 in) wide, and four thick, "strap-like" oral arms extending up to 6 metres (20 ft) in length, very different from the typical fine, threadlike tentacles that rim the umbrella of more-typical-looking jellyfish, including the Lion's Mane.

Life History and Behavior

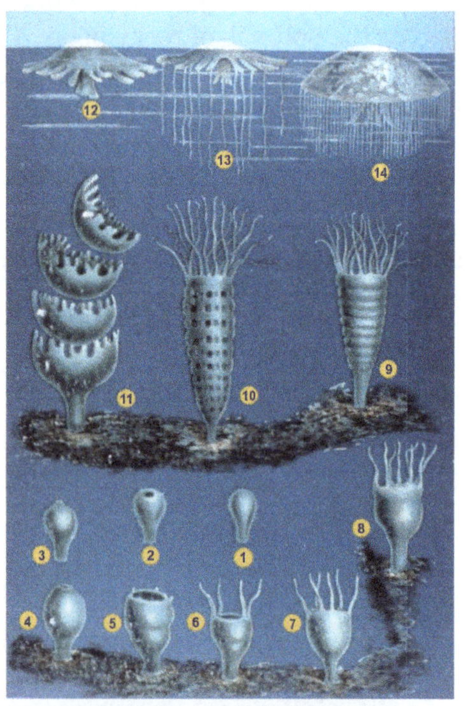

The developmental stages of scyphozoan jellyfish's life cycle: 1–3 Larva searches for site 4–8 Polyp grows 9–11 Polyp strobilates 12–14 Medusa grows

Life Cycle

Jellyfish have a complex life cycle which includes both sexual and asexual phases, with the medusa being the sexual stage in most instances. Sperm fertilize eggs, which develop into larval planulae, become polyps, bud into ephyrae and then transform into adult medusae. In some species certain stages may be skipped.

Upon reaching adult size, jellyfish spawn regularly if there is a sufficient supply of food. In most species, spawning is controlled by light, with all individuals spawning at about the same time of day, in many instances this is at dawn or dusk. Jellyfish are usually either male or female (with occasional hermaphrodites). In most cases, adults release sperm and eggs into the surrounding water, where the unprotected eggs are fertilized and develop into larvae. In a few species, the sperm swim into the female's mouth, fertilizing the eggs within her body, where they remain during early development stages. In moon jellies, the eggs lodge in pits on the oral arms, which form a temporary brood chamber for the developing planula larvae.

The planula is a small larva covered with cilia. When sufficiently developed, it settles onto a firm surface and develops into a polyp. The polyp generally consists of a small stalk topped by a mouth that is ringed by upward-facing tentacles. The polyps resemble those of closely related anthozoans, such as sea anemones and corals. The jellyfish polyp may be sessile, living on the bottom, boat hulls or other substrates, or it may be free-floating or attached to tiny bits of free-living plankton or rarely, fish or other invertebrates. Polyps may be solitary or colonial. Most polyps are only millimetres in diameter and feed continuously. The polyp stage may last for years.

After an interval and stimulated by seasonal or hormonal changes, the polyp may begin reproducing asexually by budding and, in the Scyphozoa, is called a segmenting polyp, or a scyphistoma. Budding produces more scyphistomae and also ephyrae. Budding sites vary by species; from the tentacle bulbs, the manubrium (above the mouth), or the gonads of hydromedusae. In a process known as strobilation, the polyp's tentacles are reabsorbed and the body starts to narrow, forming transverse constrictions, in several places near the upper extremity of the polyp. These deepen as the constriction sites migrate down the body, and separate segments known as ephyra detach. These are free-swimming precursors of the adult medusa stage, which is the life stage that is typically identified as a jellyfish. The ephyrae, usually only a millimeter or two across initially, swim away from the polyp and grow. Limnomedusae polyps can asexually produce a creeping *frustule* larval form, which crawls away before developing into another polyp. A few species can produce new medusae by budding directly from the medusan stage. Some hydromedusae reproduce by fission.

Lifespan

Little is known of the life histories of many jellyfish as the places on the seabed where the benthic forms of those species live have not been found. However, an asexually reproducing strobila form can sometimes live for several years, producing new medusae (ephyra larvae) each year.

An unusual species, *Turritopsis dohrnii*, formerly classified as *Turritopsis nutricula*, might be effectively immortal because of its ability under certain circumstances to transform from medusa back to the polyp stage, thereby escaping the death that typically awaits medusae post-reproduction if they have not otherwise been eaten by some other ocean organism. So far this reversal has been observed only in the laboratory.

Locomotion

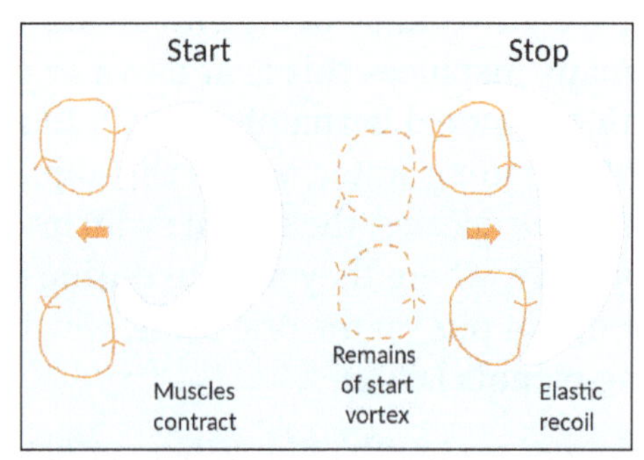

Jellyfish locomotion is highly efficient. Muscles in the jellylike bell contract, setting up a start vortex and propelling the animal. When the contraction ends, the bell recoils elastically, creating a stop vortex with no extra energy input

Using the moon jelly *Aurelia aurita* as an example, jellyfish have been shown to be the most energy efficient swimmers of all animals. They move through the water by radially expanding and contracting their bell-shaped bodies to push water behind them. They pause between the contraction and expansion phases to create two vortex rings. Muscles are used for the contraction of the body, which creates the first vortex and pushes the animal forward, but the mesoglea is so elastic that the expansion is powered exclusively by relaxing the bell, which releases the energy stored from the contraction. Meanwhile, the second vortex ring starts to spin faster, sucking water into the bell and pushing against the centre of the body, giving a secondary and "free" boost forward. The mechanism, called passive energy recapture, only works in relatively small jellyfish moving at low speeds, allowing the animal to travel 30 percent farther on each swimming cycle. Jellyfish achieved a 48 percent lower cost of transport (the amount of food and oxygen consumed, versus energy spent in movement) than other animals in similar studies. One reason for this is that most of the gelatinous tissue of the bell is inactive, using no energy during swimming.

Ecology

Diet

Jellyfish are like other cnidarians generally carnivorous (or parasitic), feeding on planktonic organisms, crustaceans, small fish, fish eggs and larvae, and other jellyfish, ingesting food and voiding undigested waste through the mouth. They hunt passively using their tentacles as drift lines, or sink through the water with their tentacles spread widely; the tentacles, which contain nematocysts to stun or kill the prey, may then flex to help bring it to the mouth. Their swimming technique also helps them to capture prey; when their bell expands it sucks in water which brings more potential prey within reach of the tentacles.

A few species such as *Aglaura hemistoma* are omnivorous, feeding on microplankton which is a mixture of zooplankton and phytoplankton (microscopic plants) such as dinoflagellates. Others harbour mutualistic algae (Zooxanthellae) in their tissues;

the spotted jellyfish (*Mastigias papua*) is typical of these, deriving part of its nutrition from the products of photosynthesis, and part from captured zooplankton.

Predation

Other species of jellyfish are among the most common and important jellyfish predators. Sea anemones may eat jellyfish that drift into their range. Other predators include tunas, sharks, swordfish, sea turtles and penguins. Jellyfish washed up on the beach are consumed by foxes, other terrestrial mammals and birds. In general however, there are few animals preying on jellyfish. Jellyfish can broadly be considered to be top predators in the food chain. Once jellyfish have become dominant in an ecosystem, for example through overfishing which removes predators of jellyfish larvae, there may be no obvious way for the previous balance to be restored: they eat fish eggs and juvenile fish, and compete with fish for food, preventing fish stocks from recovering.

Symbiosis

Some small fish are immune to the stings of the jellyfish and live among the tentacles, serving as bait in a fish trap; they are safe from potential predators and are able to share in the fish caught by the jellyfish. The cannonball jellyfish has a symbiotic relationship with ten different species of fish, and with the longnose spider crab, which lives inside the bell, sharing the jellyfish's food and nibbling its tissues.

Blooms

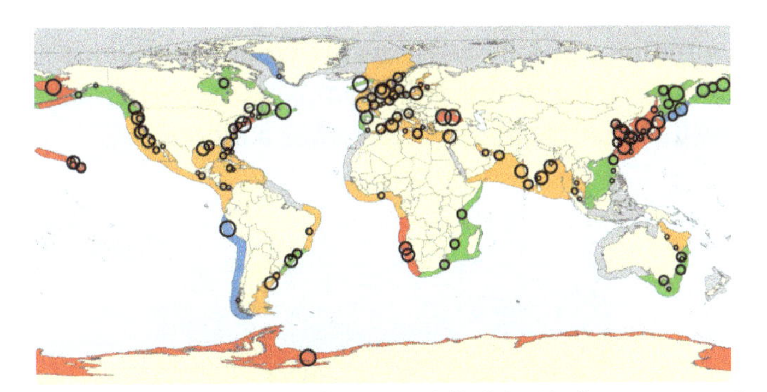

Map of population trends of native and invasive jellyfish. Circles represent data records; larger circles denote higher certainty of findings

- Increase (high certainty)
- Increase (low certainty)
- Stable/variable
- Decrease
- No data

Jellyfish form large masses or blooms in certain environmental conditions of ocean currents, nutrients, sunshine, temperature, season, prey availability, reduced predation and oxygen concentration. Currents collect jellyfish together, especially in years with unusually high populations. Jellyfish can detect marine currents and swim against the current to congregate in blooms. Jellyfish are better able to survive in nutrient-rich, oxygen-poor water than competitors, and thus can feast on plankton without competition. Jellyfish may also benefit from saltier waters, as saltier waters contain more iodine, which is necessary for polyps to turn into jellyfish. Rising sea temperatures caused by climate change may also contribute to jellyfish blooms, because many species of jellyfish are relatively better able to survive in warmer waters.

Some jellyfish populations that have shown clear increases in the past few decades are invasive species, newly arrived from other habitats: examples include the Black Sea, Caspian Sea, Baltic Sea, central and eastern Mediterranean, Hawaii, and tropical and subtropical parts of the West Atlantic (including the Caribbean, Gulf of Mexico and Brazil).

Increased nutrients from agricultural or urban runoff with nutrients including nitrogen and phosphorus compounds increase the growth of phytoplankton, causing eutrophication and algal blooms. When the phytoplankton die, they may create dead zones, so called because they are ahypoxic (low in oxygen). This in turn kills fish and other animals, but not jellyfish, allowing them to bloom. Jellyfish populations may be expanding globally as a result of land runoff and overfishing of their natural predators. Jellyfish are well placed to benefit from disturbance of marine ecosystems. They reproduce rapidly; they prey upon many species, while few species prey on them; and they feed via touch rather than visually, so they can feed effectively at night and in turbid waters. It may be difficult for fish stocks to reestablish themselves in marine ecosystems once they have become dominated by jellyfish, because jellyfish feed on plankton, which includes fish eggs and larvae.

Jellyfish form a component of jelly-falls, events where gelatinous zooplankton fall to the seafloor, providing food for the benthic organisms there. In temperate and subpolar regions, jelly-falls usually follow immediately after a bloom.

Habitats

A common Scyphozoan jellyfish seen near beaches in the Florida Panhandle

Most jellyfish are marine animals, although a few hydromedusae inhabit freshwater. The best known freshwater example is the cosmopolitan hydrozoan jellyfish, *Craspedacusta sowerbii*. It is less than an inch (2.5 cm) in diameter, colorless and does not sting. Some jellyfish populations have become restricted to coastal saltwater lakes, such as Jellyfish Lake in Palau. Jellyfish Lake is a marine lake where millions of golden jellyfish (*Mastigias* spp.) migrate horizontally across the lake daily.

Although most jellyfish live well off the ocean floor and form part of the plankton, a few species are closely associated with the bottom for much of their lives and can be considered benthic. The upside-down jellyfish in the genus *Cassiopea* typically lie on the bottom of shallow lagoons where they sometimes pulsate gently with their umbrella top facing down. Even some deep-sea species of hydromedusae and scyphomedusae are usually collected on or near the bottom. All of the stauromedusae are found attached to either seaweed or rocky or other firm material on the bottom.

Some species explicitly adapt to tidal flux. In Roscoe Bay, jellyfish ride the current at ebb tide until they hit a gravel bar, and then descend below the current. They remain in still waters until the tide rises, ascending and allowing it to sweep them back into the bay. They also actively avoid fresh water from mountain snowmelt, diving until they find enough salt.

Parasites

Jellyfish are hosts to a wide variety of parasitic organisms. They act as intermediate hosts of endoparasitic helminths, with the infection being transferred to the definitive host fish after predation. Some digenean trematodes, especially species in the family Lepocreadiidae, use jellyfish as their second intermediate hosts. Fish become infected by the trematodes when they feed on infected jellyfish.

Echinoderm

Echinoderm is the common name given to any member of the phylum Echinodermata of marine animals. The adults are recognizable by their (usually five-point) radial symmetry, and include such well-known animals as starfish, sea urchins, sand dollars, and sea cucumbers, as well as the sea lilies or "stone lilies". Echinoderms are found at every ocean depth, from the intertidal zone to the abyssal zone. The phylum contains about 7000 living species, making it the second-largest grouping of deuterostomes (a superphylum), after the chordates (which include the vertebrates, such as birds, fishes, mammals, and reptiles). Echinoderms are also the largest phylum that has no freshwater or terrestrial (land-based) representatives.

Aside from the hard-to-classify *Arkarua* (a Precambrian animal with echinoderm-like pentamerous radial symmetry), the first definitive members of the phylum appeared near the start of the Cambrian. One group of Cambrian echinoderms, the cinctans (Homalozoa), which are close to the base of the echinoderm origin, have been found to possess external gills used for filter feeding, similar to those possessed by chordates and hemichordates.

The echinoderms are important both ecologically and geologically. Ecologically, there are few other groupings so abundant in the biotic desert of the deep sea, as well as shallower oceans. Most echinoderms are able to reproduce asexually and regenerate tissue, organs, and limbs; in some cases, they can undergo complete regeneration from a single limb. Geologically, the value of echinoderms is in their ossified skeletons, which are major contributors to many limestone formations, and can provide valuable clues as to the geological environment. They were the most used species in regenerative research in the 19th and 20th centuries. Further, it is held by some scientists that the radiation of echinoderms was responsible for the Mesozoic Marine Revolution.

Anatomy and Physiology

Echinoderms evolved from animals with bilateral symmetry. Although adult echinoderms possess pentaradial, or five-sided, symmetry, echinoderm larvae are ciliated, free-swimming organisms that organize in bilateral symmetry which makes them look like embryonic chordates. Later, the left side of the body grows at the expense of the right side, which is eventually absorbed. The left side then grows in a pentaradially symmetric fashion, in which the body is arranged in five parts around a central axis. Within the Asterozoa, there can be a few exceptions from the rule. The starfish genus Leptasterias normally have six arms, although five-armed individuals can occur. Also the Brisingida have six armed species. Amongst the brittle stars, six-armed species such as Ophiothela danae, Ophiactis savignyi and Ophionotus hexactis exists, and Ophiacantha vivipara often has more than six.

Echinoderms exhibit secondary radial symmetry in portions of their body at some stage of life. This, however, is an adaptation to their sessile existence. They developed from other members of the Bilateria and exhibit bilateral symmetry in their larval stage. Many crinoids and some seastars exhibit symmetry in multiples of the basic five, with starfish such as *Labidiaster annulatus* known to possess up to fifty arms, and the sea-lily *Comaster schlegelii* having two hundred.

Skin and Skeleton

A brittle star, Ophionereis reticulata

A sea cucumber from Malaysia

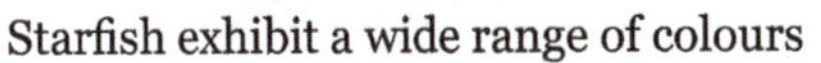

Starfish exhibit a wide range of colours

Strongylocentrotus purpuratus, a well-armoured sea urchin

Echinoderms have a mesodermal skeleton composed of calcareous plates or ossicles. Each one of these, even the articulating spine of a sea urchin, is composed mineralogically of a crystal of calcite. If solid, these would form a heavy skeleton, so they have a sponge-like porous structure known as stereom. Ossicles may be fused together, as in the test of sea urchins, or may articulate with each other as in the arms of sea stars, brittle stars and crinoids. The ossicles may be flat plates or bear external projections in the form of spines, granules or warts and they are supported by a tough epidermis (skin). Skeletal elements are also deployed in some specialized ways, such as the "Aristotle's lantern" mouthparts of sea urchins used for grinding, the supportive stalks of crinoids and the structural "lime ring" of sea cucumbers.

Despite the robustness of the individual skeletal modules complete skeletons of starfish, brittle stars and crinoids are rare in the fossil record. This is because they quickly disarticulate (disconnect from each other) once the encompassing skin rots away, and in the absence of tissue there is nothing to hold the plates together. The modular construction is a result of the growth system employed by echinoderms, which adds new segments at the centre of the radial limbs, pushing the existing plates outwards and lengthening the arms. Sea urchins on the other hand are often well preserved in chalk beds or limestone. During fossilization, the cavities in the stereom are filled in with calcite that is in crystalline continuity with the surrounding material. On fracturing such rock, distinctive cleavage patterns can be seen and sometimes even the intricate internal and external structure of the test.

The epidermis consists of cells responsible for the support and maintenance of the skeleton, as well as pigment cells, mechanoreceptor cells (which detect motion on the animal's surface), and sometimes gland cells which secrete sticky fluids or even toxins. The varied and often vivid colours of echinoderms are produced by the action of skin pigment cells. These are produced by a variable combination of coloured pigments, such as the dark melanin, red carotinoids, and carotene proteins, which can be blue, green, or violet. These may be light-sensitive, and as a result many echinoderms change appearance completely as night falls. The reaction can happen quickly – the sea urchin

Centrostephanus longispinus changes from jet black to grey-brown in just fifty minutes when exposed to light.

One characteristic of most echinoderms is a special kind of tissue known as "catch connective tissue". This collagenous material can change its mechanical properties in a few seconds or minutes through nervous control rather than by muscular means. This tissue enables a starfish to change from moving flexibly around the seabed to becoming rigid while prying open a bivalve mollusc or preventing itself from being extracted from a crevice. Similarly, sea urchins can lock their normally mobile spines rigidly as a defensive mechanism when attacked.

Water Vascular System

Echinoderms possess a unique water vascular system. This is a network of fluid-filled canals derived from the coelom (body cavity) that function in gas exchange, feeding, sensory reception and locomotion. This system varies between different classes of echinoderm but typically opens to the exterior through a sieve-like madreporite on the aboral (upper) surface of the animal. The madreporite is linked to a slender duct, the stone canal, which extends to a ring canal that encircles the mouth or oesophagus. From this, radial canals extend along the arms of asteroids and adjoin the test in the ambulacral areas of echinoids. Short lateral canals branch off the radial canals, each one ending in an ampulla. Part of the ampulla can protrude through a pore (or a pair of pores in sea urchins) to the exterior and is known as a podium or tube feet. The water vascular system assists with the distribution of nutrients throughout the animal's body and is most obviously expressed in the tube feet which can be extended or contracted by the redistribution of fluid between the foot and the internal sac.

The organization of the system is somewhat different in ophiuroids where the madreporite may be on the oral surface and the podia lack suckers. In holothuroids, the podia may be reduced or absent and the madreporite opens into the body cavity so that the circulating liquid is coelomic fluid rather than sea water. The arrangements in crinoids is similar to asteroids but the tube feet lack suckers and are used to pass food particles captured by the arms towards the central mouth. In the asteroids, the same wafting motion is employed to move the animal across the ground. Sea urchins use their feet to prevent the larvae of encrusting organisms from settling on their surfaces; potential settlers are moved to the urchin's mouth and eaten. Some burrowing sea stars extend their elongated dorsal tube feet to the surface of the sand or mud above and use them to absorb oxygen from the water column.

Other Organs

Echinoderms possess a simple digestive system which varies according to the animal's diet. Starfish are mostly carnivorous and have a mouth, oesophagus, two-part stomach, intestine and rectum, with the anus located in the centre of the aboral body surface.

With a few exceptions, the members of the order Paxillosida do not possess an anus. In many species of starfish, the large cardiac stomach can be everted and digest food outside the body. In other species, whole food items such as molluscs may be ingested. Brittle stars have a blind gut with no intestine or anus. They have varying diets and expel food waste through their mouth. Sea urchins are herbivores and use their specialised mouthparts to graze, tear and chew algae and sometimes other animal or vegetable material. They have an oesophagus, a large stomach and a rectum with the anus at the apex of the test. Sea cucumbers are mostly detritivores, sorting through the sediment with their buccal tentacles which are modified tube feet. Sand and mud accompanies their food through their simple gut which has a long coiled intestine and a capacious cloaca. Crinoids are passive suspension feeders, catching plankton with their outstretched arms. Boluses of mucus-trapped food are passed to the mouth which is linked to the anus by a loop consisting of a short oesophagus and longer intestine.

The coelomic cavities of echinoderms are complex. Aside from the water vascular system, echinoderms have a haemal coelom (or haemal system, the "haemal" being a misnomer), a perivisceral coelom, a gonadal coelom and often also a perihaemal coelom (or perihaemal system). During development, echinoderm coelom is divided in metacoel, mesocoel and protocoel (also called somatocoel, hydrocoel and axocoel, respectively). The water vascular system, haemal system and perihaemal system form the tubular coelomic system. Echinoderms are an exception having both a coelomic circulatory system (i.e., the water vascular system) and a haemal circulatory system (i.e., the haemal and perihaemal systems).

Haemal and perihaemal systems are derived from the coelom and form an open and reduced circulatory system. This usually consists of a central ring and five radial vessels. There is no true heart and the blood often lacks any respiratory pigment. Gaseous exchange occurs via dermal branchae or papulae in starfish, genital bursae in brittle stars, peristominal gills in sea urchins and cloacal trees in sea cucumbers. Exchange of gases also takes place through the tube feet. Echinoderms lack specialized excretory (waste disposal) organs and so nitrogenous waste, chiefly in the form of ammonia, diffuses out through the respiratory surfaces.

The coelomic fluid contains the coelomocytes, or immune cells. There are several types of immune cells, which vary among classes and species. All classes possess a type of phagocytic amebocyte, which engulf invading particles and infected cells, aggregate or clot, and may be involved in cytotoxicity. These cells are usually larger and granular, and are suggested to be a main line of defense against potential pathogens. Depending on the class, echinoderms may have spherule cells (for cytotoxicity, inflammation, and anti-bacterial activity), vibratile cells (for coelomic fluid movement and clotting), and crystal cells (potential osmoregulatory cells in sea cucumbers). The coelomocytes also secrete Anti-Microbial Peptides (AMPs) against bacteria, and have a set of lectins and complement proteins as part of an innate immune system that is still being characterized.

Echinoderms have a simple radial nervous system that consists of a modified nerve net consisting of interconnecting neurons with no central brain, although some do possess ganglia. Nerves radiate from central rings around the mouth into each arm or along the body wall; the branches of these nerves coordinate the movements of the organism and the synchronisation of the tube feet. Starfish have sensory cells in the epithelium and have simple eyespots and touch-sensitive tentacle-like tube feet at the tips of their arms. Sea urchins have no particular sense organs but do have statocysts that assist in gravitational orientation, and they have sensory cells in their epidermis, particularly in the tube feet, spines and pedicellariae. Brittle stars, crinoids and sea cucumbers in general do not have sensory organs but some burrowing sea cucumbers of the order Apodida have a single statocyst adjoining each radial nerve and some have an eyespot at the base of each tentacle.

The gonads occupy much of the body cavities of sea urchins and sea cucumbers, while the less voluminous crinoids, brittle stars and starfish have two gonads in each arm. While the ancestral condition is considered to be the possession of one genital aperture, many organisms have multiple gonopores through which eggs or sperm may be released.

Regeneration

Sunflower star regenerating several arms

Many echinoderms have remarkable powers of regeneration. Many species routinely autotomize and regenerate arms and viscera. Sea cucumbers often discharge parts of their internal organs if they perceive themselves to be threatened. The discharged organs and tissues are regenerated over the course of several months. Sea urchins are constantly replacing spines lost through damage. Sea stars and sea lilies readily lose and regenerate their arms. In most cases, a single severed arm cannot grow into a new starfish in the absence of at least part of the disc. However, in a few species a single arm can survive and develop into a complete individual and in some species, the arms are intentionally detached for the purpose of asexual reproduction. During periods when they have lost their digestive tracts, sea cucumbers live off stored nutrients and absorb dissolved organic matter directly from the water.

The regeneration of lost parts involves both epimorphosis and morphallaxis. In epimorphosis stem cells—either from a reserve pool or those produced by dedifferentiation—form a blastema and generate new tissues. Morphallactic regeneration involves the movement and remodelling of existing tissues to replace lost parts. Direct transdifferentiation of one type of tissue to another during tissue replacement is also observed.

The robust larval growth is responsible for many echinoderms being used as popular model organisms in developmental biology.

Distribution and Habitat

Echinoderms are globally distributed in almost all depths, latitudes and environments in the ocean. They reach highest diversity in reef environments but are also widespread on shallow shores, around the poles — refugia where crinoids are at their most abundant — and throughout the deep ocean, where bottom-dwelling and burrowing sea cucumbers are common — sometimes accounting for up to 90% of organisms. While almost all echinoderms are benthic — that is, they live on the sea floor — some sea-lilies can swim at great velocity for brief periods of time, and a few deep-sea sea cucumbers are fully floating. Some crinoids are pseudo-planktonic, attaching themselves to floating logs and debris, although this behaviour was exercised most extensively in the Paleozoic, before competition from such organisms as barnacles restricted the extent of the behaviour.

The larvae of echinoderms, especially starfish and sea urchins, are pelagic, and with the aid of ocean currents can be transported for great distances, reinforcing the global distribution of the phylum.

Mode of Life

Locomotion

Echinoderms primarily use their tube feet to move about but some sea urchins also use their spines. The tube feet typically have a tip shaped like a suction pad in which a vacuum can be created by contraction of muscles. This along with some stickiness provided by the secretion of mucus provides adhesion. Waves of tube feet contractions and relaxations move along the adherent surface and the animal moves slowly along.

Brittle stars are the most agile of the echinoderms, raising their discs and taking strides when moving. The two forward arms grip the substrate with their tube feet, the two side arms "row", the hindermost arm trails and the animal moves in jerks. The arm spines provide traction and when moving among objects, the supple arms can coil around things. A few species creep around on pointed tube feet. Starfish extend their tube feet in the intended direction of travel and grip the substrate by suction, after which the feet are drawn backwards. The movement of multiple tube feet, coordinated in waves, moves the animal forward, but progress is slow. Some burrowing starfish have points rather than suckers on their tube feet and they are able to "glide" across the seabed at a faster rate.

Sea urchins use their tube feet to move around in a similar way to starfish. Some also use their articulated spines to push or lever themselves along or lift their oral surfaces off the substrate. If a sea urchin is overturned, it can extend its tube feet in one ambulacral area far enough to bring them within reach of the substrate and then successively attach feet from the adjoining area until it is righted. Some species bore into rock and they usually do this by grinding away at the surface with their mouthparts.

Sea cucumbers are generally sluggish animals. Many can move on the surface or burrow through sand or mud using peristaltic movements and some have short tube feet on their under surface with which they can creep along in the manner of a starfish. Some species drag themselves along by means of their buccal tentacles while others can expand and contract their body or rhythmically flex it and "swim". Many live in cracks, hollows and burrows and hardly move at all. Some deep water species are pelagic and can float in the water with webbed papillae forming sails or fins.

The majority of crinoids are motile but the sea lilies are sessile and attached to hard substrates by stalks. These stems can bend and the arms can roll and unroll and that is about the limit of the sea lily's movement, although a few species can relocate themselves on the seabed by crawling. The sea feathers are unattached and usually live in crevices, under corals or inside sponges with their arms the only visible part. Some sea feathers emerge at night and perch themselves on nearby eminences to better exploit the food-bearing current. Many species can "walk" across the seabed, raising their body with the help of their arms. Many can also swim with their arms but most are largely sedentary, seldom moving far from their chosen place of concealment.

Feeding

The modes of feeding vary greatly between the different echinoderm taxa. Crinoids and some brittle stars tend to be passive filter-feeders, enmeshing suspended particles from passing water; most sea urchins are grazers, sea cucumbers deposit feeders and the majority of starfish are active hunters.

Crinoids are suspension feeders and spread their arms wide to catch particles floating past. These are caught by the tube feet on the pinnules, moved into the ambulacral grooves, wrapped in mucus and conveyed to the mouth by the cilia lining the grooves. The exact dietary requirements of crinoids have been little researched but in the laboratory they can be fed with diatoms.

Basket stars are suspension feeders, raising their branched arms to collect zooplankton, while brittle stars use several methods of feeding, though usually one predominates. Some are suspension feeders, securing food particles with mucus strands, spines or tube feet on their raised arms. Others are scavengers and feeders on detritus. Others again are voracious carnivores and able to lasso their waterborne prey with a sudden encirclement by their flexible arms. The limbs then bend under the disc to transfer the food to the jaws and mouth.

Many sea urchins feed on algae, often scraping off the thin layer of algae covering the surfaces of rocks with their specialised mouthparts known as Aristotle's lantern. Other species devour smaller organisms, which they may catch with their tube feet. They may also feed on dead fish and other animal matter. Sand dollars may perform suspension feeding and feed on phytoplankton, detritus, algal pieces and the bacterial layer surrounding grains of sand.

Many sea cucumbers are mobile deposit or suspension feeders, using their buccal podia to actively capture food and then stuffing the particles individually into their buccal cavities. Others ingest large quantities of sediment, absorb the organic matter and pass the indigestible mineral particles through their guts. In this way they disturb and process large volumes of substrate, often leaving characteristic ridges of sediment on the seabed. Some sea cucumbers live infaunally in burrows, anterior-end down and anus on the surface, swallowing sediment and passing it through their gut. Other burrowers live anterior-end up and wait for detritus to fall into the entrances of the burrows or rake in debris from the surface nearby with their buccal podia.

Nearly all starfish are detritivores or carnivores, though a few are suspension feeders. Small fish landing on the upper surface may be captured by pedicilaria and dead animal matter may be scavenged but the main prey items are living invertebrates, mostly bivalve molluscs. To feed on one of these, the starfish moves over it, attaches its tube feet and exerts pressure on the valves by arching its back. When a small gap between the valves is formed, the starfish inserts part of its stomach into the prey, excretes digestive enzymes and slowly liquefies the soft body parts. As the adductor muscle of the shellfish relaxes, more stomach is inserted and when digestion is complete, the stomach is returned to its usual position in the starfish with its now liquefied bivalve meal inside it. The same everted stomach process is used by other starfish to feed on sponges, sea anemones, corals, detritus and algal films.

Defense Mechanisms

Despite their low nutrition value and the abundance of indigestible calcite, echinoderms are the prey of many organisms, such as crabs, sharks, sea birds and other echinoderms. Defensive strategies employed include the presence of spines, toxins, which can be inherent or delivered through the tube feet, and the discharge of sticky entangling threads by sea cucumbers. Although most echinoderm spines are blunt, those of the crown-of-thorns starfish are long and sharp and can cause a painful puncture wound as the epithelium covering them contains a toxin. Because of their catch connective tissue, which can change rapidly from a flaccid to a rigid state, echinoderms are very difficult to dislodge from crevices. Certain sea cucumbers have a cluster of cuvierian tubules which can be ejected as long sticky threads from their anus and entangle and permanently disable an attacker. Another defensive strategy sometimes adopted by sea cucumbers is to rupture the body wall and discharge the gut and internal organs. The animal has a great regenerative capacity and will regrow the lost parts later. Starfish and brittle stars may undergo autotomy when attacked, an arm becoming detached which may distract the predator for long enough for the animal to escape. Some starfish species can "swim" away from what may be danger, foregoing the regrowth by not losing limbs. It is not unusual to find starfish with arms of different sizes in various stages of regrowth.

Ecology

Echinoderms are numerous and relatively large invertebrates and play an important role in marine, benthic ecosystems. The grazing of sea urchins reduces the rate of colonization of bare rock by settling organisms but also keeps algae in check, thereby enhancing the biodiversity of coral reefs. The burrowing of sand dollars, sea cucumbers and some starfish stirs up the sediment and depletes the sea floor of nutrients. Their digging activities increases the depth to which oxygen can seep and allows a more complex ecological tier-system to develop. Starfish and brittle stars prevent the growth of algal mats on coral reefs, which might otherwise obstruct the filter-feeding constituent organisms. Some sea urchins can bore into solid rock and this bioerosion can destabilise rock faces and release nutrients into the ocean. Coral reefs are also bored into in this way but the rate of accretion of carbonate material is often greater than the erosion produced by the sea urchin. It has been estimated that echinoderms capture and sequester about 0.1 gigatonnes of carbon per year as calcium carbonate, making them important contributors in the global carbon cycle.

Echinoderms sometimes have large population swings which can cause marked consequences for ecosystems. An example is the change from a coral-dominated reef system to an alga-dominated one that resulted from the mass mortality of the tropical sea urchin *Diadema antillarum* in the Caribbean in 1983. Sea urchins are among the main herbivores on reefs and there is usually a fine balance between the urchins and the kelp and other algae on which they graze. A diminution of the numbers of predators (otters, lobsters and fish) can result in an increase in urchin numbers causing overgrazing of kelp forests with the result that an alga-denuded "urchin barren" forms. On the Great Barrier Reef, an unexplained increase in the numbers of crown-of-thorns starfish (*Acanthaster planci*), which graze on living coral tissue, has had considerable impact on coral mortality and coral reef biodiversity.

Echinoderms form part of the diet of many organisms such as bony fish, sharks, eider ducks, gulls, crabs, gastropod molluscs, sea otters, Arctic foxes and humans. Larger starfish prey on smaller ones and the great quantity of eggs and larvae produced form part of the zooplankton, consumed by many marine creatures. Crinoids are relatively free from predation. The body cavities of many sea cucumbers and some starfish provide a habitat for parasitic or symbiotic organisms including fish, crabs, worms and snails.

Use by Humans

In 2010, 373,000 tonnes of echinoderms were harvested, mainly for consumption. These were mainly sea cucumbers (158,000 tonnes) and sea urchins (73,000 tonnes).

Sea cucumbers are considered a delicacy in some countries of south east Asia; as such, they are in imminent danger of being over-harvested.

Popular species include the pineapple roller *Thelenota ananas* (*susuhan*) and the red *Holothuria edulis*. These and other species are colloquially known as *bêche de mer* or *trepang* in China and Indonesia. The sea cucumbers are boiled for twenty minutes and then dried both naturally and later over a fire which gives them a smoky tang. In China they are used as a basis for gelatinous soups and stews. Both male and female gonads of sea urchins are also consumed particularly in Japan, Peru, Spain and France. The taste is described as soft and melting, like a mixture of seafood and fruit. The quality is assessed by the colour which can range from light yellow to bright orange. At the present time, some trials of breeding sea uchins in order to try to compensate the overexploitation of this resource have been made.

The calcareous tests or shells of echinoderms are used as a source of lime by farmers in areas where limestone is unavailable and some are used in the manufacture of fish meal. Four thousand tons of the animals are used annually for these purposes. This trade is often carried out in conjunction with shellfish farmers, for whom the starfish pose a major threat by eating their cultured stock. Other uses for the starfish they recover include the manufacture of animal feed, composting and drying for the arts and craft trade.

Sea urchins are used in research, particularly as model organisms in developmental biology and ecotoxicology. *Strongylocentrotus purpuratus* and *Arbacia punctulata* are used for this purpose in embryological studies. The large size and the transparency of the eggs enables the observation of sperm cells in the process of fertilising ova. The arm regeneration potential of brittle stars is being studied in connection with understanding and treating neurodegenerative diseases in humans.

Sea Cucumber

Sea cucumbers are echinoderms from the class Holothuroidea. They are marine animals with a leathery skin and an elongated body containing a single, branched gonad. Sea cucumbers are found on the sea floor worldwide. The number of holothurian species worldwide is about 1,717 with the greatest number being in the Asia Pacific region. Many of these are gathered for human consumption and some species are cultivated in aquaculture systems. The harvested product is variously referred to as *trepang*, *namako*, *bêche-de-mer* or *balate*. Sea cucumbers serve a useful role in the marine ecosystem as they help recycle nutrients, breaking down detritus and other organic matter after which bacteria can continue the degradation process.

Like all echinoderms, sea cucumbers have an endoskeleton just below the skin, calcified structures that are usually reduced to isolated microscopic ossicles (or sclerietes) joined by connective tissue. In some species these can sometimes be enlarged to flattened plates, forming an armour. In pelagic species such as *Pelagothuria natatrix* (Order Elasipodida, family Pelagothuriidae), the skeleton is absent and there is no calcareous ring.

The sea cucumbers are named after their resemblance to the fruit of the cucumber plant.

Details of the mouth with its tentacles

Synaptula lamperti lives on sponges (here in Indonesia)

Synapta maculata, the longest known sea cucumber (Apodida)

The king sea cucumber (Thelenota anax, family Stichopodidae) is one of the heaviest known holothurians

Most sea cucumbers, as their name suggests, have a soft and cylindrical body, more or less lengthened, rounded off and occasionally fat in the extremities, and generally without solid appendages. Their shape ranges from almost spherical for "sea apples" (genus *Pseudocolochirus*) to serpent-like for Apodida or the classic sausage-shape, while others resemble caterpillars. The mouth is surrounded by tentacles, which can be pulled back inside the animal. Holothurians measure generally between 10 and 30 centimetres long, with extremes of some millimetres for *Rhabdomolgus ruber* and up to more than 3 metres for *Synapta maculata*. The largest American species, *Holothuria floridana*, which abounds just below low-water mark on the Florida reefs, has a volume of well over 500 cubic centimeters (31 cu in), and 25–30 cm (10–12 in) long. Most possess five rows of tube feet (called "podia"), but Apodida lacks these and moves by crawling; the podia can be of smooth aspect or provided with fleshy appendages (like *Thelenota ananas*). The podia on the dorsal surface generally have no locomotive role, and are transformed into papillae. At one of the extremities opens a rounded mouth, generally surrounded with a crown of tentacles which can be very complex in some species (they are in fact modified podia); the anus is postero-dorsal.

Holothurians do not look like other echinoderms at first glance, because of their tubular body, without visible skeleton nor hard appendixes. Furthermore, the fivefold symmetry, classical for echinoderms, although preserved structurally, is doubled here by a bilateral symmetry which makes them look like chordates. However, a central symmetry is still visible in some species through five 'radii', which extend from the mouth to the anus, on which the tube feet are attached. There is thus no "oral" or "aboral" face as for sea stars and other echinoderms, but the animal stands on one of its sides, and this face is called *trivium* (with three rows of tube feet), while the dorsal face is named *bivium*. A remarkable feature of these animals is the "catch" collagen that forms their body wall. This can be loosened and tightened at will, and if the animal wants to squeeze through a small gap, it can essentially liquefy its body and pour into the space. To keep itself safe in these crevices and cracks, the sea cucumber will hook up all its collagen fibers to make its body firm again.

Chiridota heheva, abyssal species

Cucumaria miniata, a filter-feeding sea cucumber

Pseudocolochirus ("sea apple")

Holothuria leucospilota

The most common way to separate the subclasses is by looking at their oral tentacles. Order Apodida have a slender and elongate body lacking tube feet, with up to 25 simple or pinnate oral tentacles. Aspidochirotida are the most common sea cucumbers encountered, with a strong body and 10–30 leaf like or shield like oral tentacles. Dendrochirotida are filter-feeders, with plump bodies and 8–30 branched oral tentacles (which can be extremely long and complex).

Isostichopus badionotus

Thelenota rubralineata

Holothuria fuscopunctata

Bohadschia argus

Anatomy

Sea cucumbers are typically 10 to 30 cm (4 to 12 in) in length, although the smallest known species are just 3 mm (0.12 in) long, and the largest can reach 3 meters (10 ft). The body ranges from almost spherical to worm-like, and lacks the arms found in many other echinoderms, such as starfish. The anterior end of the animal, containing the mouth, corresponds to the oral pole of other echinoderms (which, in most cases, is the underside), while the posterior end, containing the anus, corresponds to the aboral pole. Thus, compared with other echinoderms, sea cucumbers can be said to be lying on their side.

Conspicuous Sea Cucumber, Coconut Island, Hawaii

Body Plan

The body of a holothurian is roughly cylindrical. It is radially symmetrical along its longitudinal axis, and has weak bilateral symmetry transversely with a dorsal and a ventral surface. As in other Echinozoans, there are five ambulacra separated by five ambulacral grooves, the interambulacra. The ambulacral grooves bear four rows of tube feet but these are diminished in size or absent in some holothurians, especially on the dorsal surface. The two dorsal ambulacra make up the bivium while the three ventral ones are known as the trivium.

At the anterior end, the mouth is surrounded by a ring of tentacles which are usually retractable into the mouth. These are modified tube feet and may be simple, branched or arborescent. They are known as the introvert and posterior to them there is an internal ring of large calcareous ossicles. Attached to this are five bands of muscle running internally longitudinally along the ambulacra. There are also circular muscles, contraction of which cause the animal to elongate and the introvert to extend. Anterior to the ossicles lie further muscles, contraction of which cause the introvert to retract.

The body wall consists of an epidermis and a dermis and contains smaller calcareous ossicles, the types of which are characteristics which help to identify different species. Inside the body wall is the coelom which is divided by three longitudinal mesenteries which surround and support the internal organs.

Digestive System

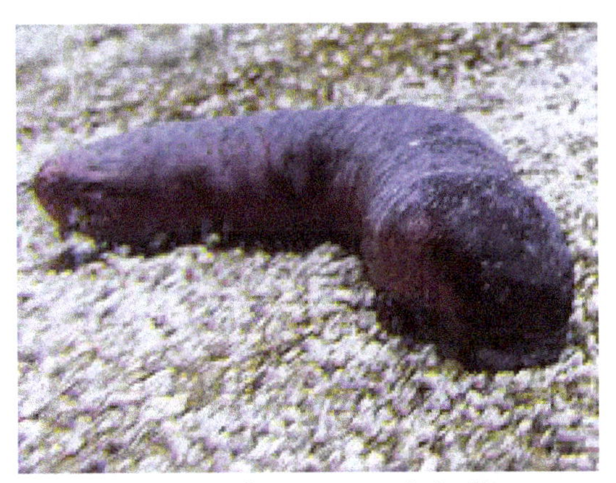

A sea cucumber atop gravel, feeding

A pharynx lies behind the mouth and is surrounded by a ring of ten calcareous plates. In most sea cucumbers, this is the only substantial part of the skeleton, and it forms the point of attachment for muscles that can retract the tentacles into the body for safety as for the main muscles of the body wall. Many species possess an oesophagus and stomach, but in some the pharynx opens directly into the intestine. The intestine is typically long and coiled, and loops through the body three times before terminating in a cloacal chamber, or directly as the anus.

Nervous System

Sea cucumbers have no true brain. A ring of neural tissue surrounds the oral cavity, and sends nerves to the tentacles and the pharynx. The animal is, however, quite capable of functioning and moving about if the nerve ring is surgically removed, demonstrating that it does not have a central role in nervous coordination. In addition, five major nerves run from the nerve ring down the length of the body beneath each of the ambulacral areas.

Most sea cucumbers have no distinct sensory organs, although there are various nerve endings scattered through the skin, giving the animal a sense of touch and a sensitivity to the presence of light. There are, however, a few exceptions: members of the Apodida order are known to possess statocysts, while some species possess small eye-spots near the bases of their tentacles.

Respiratory System

Sea cucumbers extract oxygen from water in a pair of "respiratory trees" that branch in the cloaca just inside the anus, so that they "breathe" by drawing water in through the anus and then expelling it. The trees consist of a series of narrow tubules branching from a common duct, and lie on either side of the digestive tract. Gas exchange occurs across the thin walls of the tubules, to and from the fluid of the main body cavity.

Together with the intestine, the respiratory trees also act as excretory organs, with nitrogenous waste diffusing across the tubule walls in the form of ammonia and phagocytic coelomocytes depositing particulate waste.

Circulatory Systems

Like all echinoderms, sea cucumbers possess both a water vascular system that provides hydraulic pressure to the tentacles and tube feet, allowing them to move, and a *haemal system*. The latter is more complex than that in other echinoderms, and consists of well-developed vessels as well as open sinuses.

A central haemal ring surrounds the pharynx next to the ring canal of the water vascular system, and sends off additional vessels along the radial canals beneath the ambulacral areas. In the larger species, additional vessels run above and below the intestine and are connected by over a hundred small muscular ampullae, acting as miniature hearts to pump blood around the haemal system. Additional vessels surround the respiratory trees, although they contact them only indirectly, via the coelomic fluid.

Indeed, the blood itself is essentially identical with the coelomic fluid that bathes the organs directly, and also fills the water vascular system. Phagocytic coelomocytes, somewhat similar in function to the white blood cells of vertebrates, are formed within the haemal vessels, and travel throughout the body cavity as well as both circulatory systems. An additional form of coelomocyte, not found in other echinoderms, has a flattened discoid shape, and contains hemoglobin. As a result, in many (though not all) species, both the blood and the coelomic fluid are red in colour.

Pearsonothuria graeffei showing its three rows of podia on its trivium

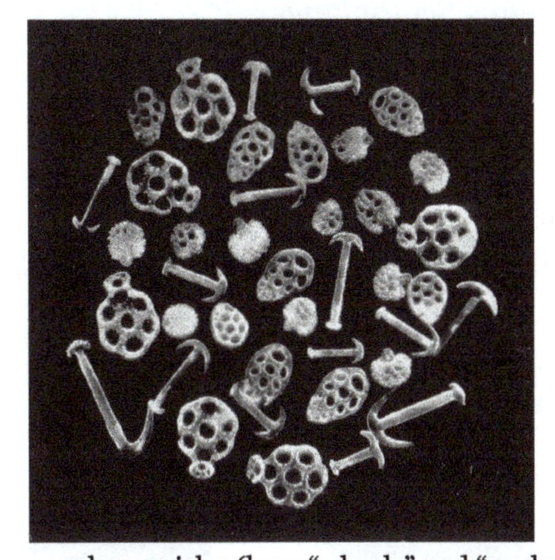

Sea cucumber ossicles (here "wheels" and "anchors")

Vanadium has been reported in high concentrations in holothurian blood, however researchers have been unable to reproduce these results.

Locomotive Organs

Like all echinoderms, sea cucumbers possess pentaradial symmetry, with their bodies divided into five nearly identical parts around a central axis. However, because of their posture, they have secondarily evolved a degree of bilateral symmetry. For example, because one side of the body is typically pressed against the substratum, and the other is not, there is usually some difference between the two surfaces (except for Apodida). Like sea urchins, most sea cucumbers have five strip-like ambulacral areas running along the length of the body from the mouth to the anus. The three on the lower surface have numerous tube feet, often with suckers, that allow the animal to crawl along; they are called *trivium*. The two on the upper surface have under-developed or vestigial tube feet, and some species lack tube feet altogether; this face is called *bivium*.

In some species, the ambulacral areas can no longer be distinguished, with tube feet spread over a much wider area of the body. Those of the order Apodida have no tube feet or ambulacral areas at all, and burrow through sediment with muscular contractions of their body similar to that of worms, however five radial lines are generally still obvious along their body.

Even in those sea cucumbers that lack regular tube feet, those that are immediately around the mouth are always present. These are highly modified into retractile tentacles, much larger than the locomotive tube feet. Depending on the species, sea cucumbers have between ten and thirty such tentacles and these can have a wide variety of shapes depending on the diet of the animal and other conditions.

Many sea cucumbers have papillae, conical fleshy projections of the body wall with sensory tube feet at their apices. These can even evolve into long antennae-like structures, especially on the abyssal genus *Scotoplanes*.

Endoskeleton

Echinoderms typically possess an internal skeleton composed of plates of calcium carbonate. In most sea cucumbers, however, these have become reduced to microscopic ossicles embedded beneath the skin. A few genera, such as *Sphaerothuria*, retain relatively large plates, giving them a scaly armour.

Life History and Behaviour

Habitat

The mysterious Pelagothuria natatrix is the only truly pelagic echinoderm known to date

Spanish dancer (Benthodytes sp.), another swimming sea cucumber, hovering at 2789 meters by the Davidson Seamount

Sea cucumbers can be found in great numbers on the deep seafloor, where they often make up the majority of the animal biomass. At depths deeper than 8.9 km (5.5 mi), sea cucumbers comprise 90% of the total mass of the macrofauna. Sea cucumbers form large herds that move across the bathygraphic features of the ocean, hunting food. The body of some deep water holothurians, such as *Enypniastes eximia*, *Peniagone leander* and *Paelopatides confundens*, is made of a tough gelatinous tissue with unique properties that makes the animals able to control their own buoyancy, making it possible for them to either live on the ocean floor or to actively swim or float over it in order to move to new locations, in a manner similar to how the group Torquaratoridae floats through water.

Benthopelagic sea cucumbers, such as this Enypniastes, are often confused with jellyfish, have webbed swimming structures enabling them to swim up off the surface of the seafloor and journey as much as 1,000 m (3,300 ft) up the water column

Holothurians appear to be the echinoderms best adapted to extreme depths, and are still very diversified beyond 5,000 m deep: several species from the family Elpidiidae ("sea pigs") can be found deeper than 9,500 m, and the record seems to be some species of the genus *Myriotrochus* (in particular *Myriotrochus bruuni*), identified down to 10,687 meters deep. In more shallow waters, sea cucumbers can form dense populations. The strawberry sea cucumber (*Squamocnus brevidentis*) of New Zealand lives on rocky walls around the southern coast of the South Island where populations sometimes reach densities of 1,000 animals per square meter (93 /sq ft). For this reason, one such area in Fiordland is called the strawberry fields.

Locomotion

Some abyssal species in the abyssal order Elasipodida have evolved to a "benthopelagic" behaviour: their body is nearly the same density as the water around them, so they can make long jumps (up to 1,000 m (3,300 ft) high), before falling slowly back to the ocean floor. Most of them have specific swimming appendages, such as some kind of umbrella (like *Enypniastes*), or a long lobe on top of the body (*Psychropotes*). Only one species is known as a true completely pelagic species, that never comes close to the bottom: *Pelagothuria natatrix*.

Diet

Holothuroidea are generally scavengers, feeding on debris in the benthic zone of the ocean. Exceptions include some pelagic cucumbers and the species *Rynkatorpa pawsoni*, which has a commensal relationship with deep-sea anglerfish. The diet of most cucumbers consists of plankton and decaying organic matter found in the sea. Some sea cucumbers position themselves in currents and catch food that flows by with their open tentacles. They also sift through the bottom sediments using their tentacles. Other species can dig into bottom silt or sand until they are completely buried. They then extrude their feeding tentacles, ready to withdraw at any hint of danger.

The mouth of an Euapta godeffroyi, showing pinnate tentacles

Mouth of a Holothuria sp., showing peltate tentacles

In the South Pacific sea cucumbers may be found in densities of 40 individuals per square meter (33/sq yd). These populations can process 19 kilograms of sediment per square meter (34 lb/sq yd) per year.

Mouth of a Cucumaria miniata, with dendritic tentacles, for filtering the water

Faeces of an holothurian. This participates in sediment homogenization and purification

The shape of the tentacles is generally adapted to the diet, and to the size of the particles to be ingested: the filter-feeding species mostly have complex arborescent tentacles, intended to maximize the surface area available for filtering, while the species feeding on the substratum will more often need digitate tentacles to sort out the nutritional material; the detritivore species living on fine sand or mud more often need shorter "peltate" tentacles, shaped like shovels. A single specimen can swallow more than 45 kg of sediment a year, and their excellent digestive capacities allow them to reject a finer, purer and homogeneous sediment. Therefore, sea cucumbers play a major role in the biological processing of the sea bed (bioturbation, purge, homogenization of the substratum etc.).

Communication and Sociability

Reproduction

Most sea cucumbers reproduce by releasing sperm and ova into the ocean water. Depending on conditions, one organism can produce thousands of gametes. Sea cucumbers are typically dioecious, with separate male and female individuals, but some species are protandric. The reproductive system consists of a single gonad, consisting of a cluster of tubules emptying into a single duct that opens on the upper surface of the animal, close to the tentacles.

"Auricularia" larva(by Ernst Haeckel)

At least 30 species, including the red-chested sea cucumber (*Pseudocnella insolens*), fertilize their eggs internally and then pick up the fertilized zygote with one of their feeding tentacles. The egg is then inserted into a pouch on the adult's body, where it develops and eventually hatches from the pouch as a juvenile sea cucumber. A few species are known to brood their young inside the body cavity, giving birth through a small rupture in the body wall close to the anus.

Development

In all other species, the egg develops into a free-swimming larva, typically after around three days of development. The first stage of larval development is known as an auricularia, and is only around 1 mm (39 mils) in length. This larva swims by means of a long band of cilia wrapped around its body, and somewhat resembles the bipinnaria larva of starfish. As the larva grows it transforms into the doliolaria, with a barrel-shaped body and three to five separate rings of cilia. The "pentacularia" is the third larval stage of sea cucumber, where the tentacles appear. The tentacles are usually the first adult features to appear, before the regular tube feet.

Symbiosis and Commensalism

Numerous small animals can live in symbiosis or commensalism with sea cucumbers, as well as some parasites.

Emperor shrimp Periclimenes imperator on a Bohadschia ocellata sea cucumber

Some cleaner shrimps can live on the tegument of holothurians, in particular several species of the genus *Periclimenes* (genus which is specialized in echinoderms), in particular *Periclimenes imperator*. A variety of fish, most commonly pearl fish, have evolved a commensalistic symbiotic relationship with sea cucumbers in which the pearl fish will live in sea cucumber's cloaca using it for protection from predation, a source of food (the nutrients passing in and out of the anus from the water), and to develop into their adult stage of life. Many polychaete worms (family Polynoidae) and crabs (like *Lissocarcinus orbicularis*) have also specialized to use the mouth or the cloacal respiratory trees for protection by living inside the sea cucumber. Nevertheless, holothurians species of the genus *Actinopyga* have anal teeth that prevent visitors from penetrating their anus.

Sea cucumbers can also shelter bivalvia as endocommensals, such as *Entovalva sp.*

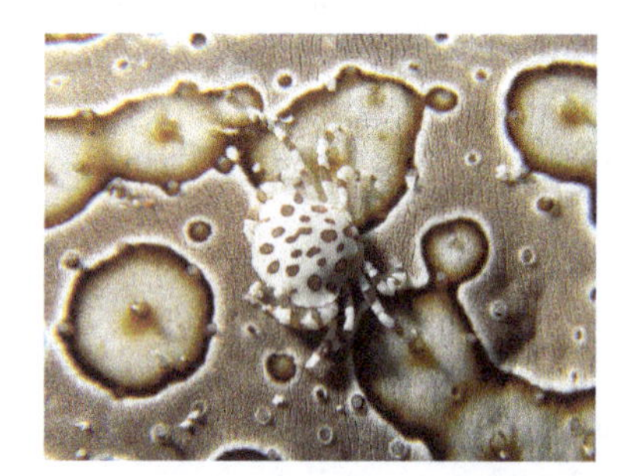

Lissocarcinus orbicularis, a symbiotic crab

Periclimenes imperator, a symbiotic shrimp

Polynoid worms on a king sea cucumber

Predators and Defensive Systems

Sea cucumbers are often scorned by most of the marine predators because of the toxins they contain (in particular holothurin) and because of their often spectacular defensive systems. However, they remain a prey for some highly specialized predators which are not affected by their toxins, such as the big mollusks *Tonna galea* and *Tonna perdix*, which paralyzes them using powerful poison before swallowing them completely. Some other less specialized and opportunist predators can also prey on sea cucumbers sometimes when they cannot find any better food, such as certain species of fish (triggerfish, pufferfish) and crustaceans (crabs, lobsters, hermit crabs).

Tonna perdix, a selective predator of tropical sea cucumbers

A sea cucumber in Mahé, Seychelles ejects sticky filaments from the anus in self-defense

Some species of coral-reef sea cucumbers within the order Aspidochirotida can defend themselves by expelling their sticky cuvierian tubules (enlargements of the respiratory tree that float freely in the coelom) to entangle potential predators. When startled, these cucumbers may expel some of them through a tear in the wall of the cloaca in an autotomic process known as evisceration. Replacement tubules grow back in one and a half to five weeks, depending on the species. The release of these tubules can also be accompanied by the discharge of a toxic chemical known as holothurin, which has similar properties to soap. This chemical can kill animals in the vicinity and is one more method by which these sedentary animals can defend themselves.

Estivation

If the water temperature becomes too high, some species of sea cucumber from temperate seas can aestivate. While they are in this state of dormancy they stop feeding, their gut atrophies, their metabolism slows down and they lose weight. The body returns to its normal state when conditions improve.

Sea Urchin

Sea urchins are typically small, rounded, spiny creatures found on shallow rocky marine coastlines. The primary hazard associated with sea urchins is contact with their spines.

Biology and Identification

Sea urchins are echinoderms, a phylum of marine animals shared with starfish, sand dollars and sea cucumbers. Echinoderms are recognizable because of their pentaradial symmetry (they have five rays of symmetry), which is easily observed on a starfish. This symmetry corresponds with a water vascular system used for locomotion, transportation of nutrients and waste, and respiration. Sea urchins have tubular feet called pedicellariae, which enable movement. In one genus of sea urchin — the flower sea urchin — some of the pedicellariae have evolved into toxic claws. In this species, the spines are short and harmless, but these toxic claws can inflict an envenomation.

Sea urchins feed on organic matter in the seabed. Their mouth is located on the base of their shell, and their anus is on the top. The color of sea urchins varies depending on the species — shades of black, red, brown, green, yellow and pink are common.

Distribution

There are species of sea urchins in all oceans, from tropical to arctic waters. Most incidents between humans and sea urchins occur in tropical and subtropical waters.

Mechanism of Injury

Sea urchins are covered in spines, which can easily penetrate divers' boots and wetsuits, puncture the skin and break off. These spines are made of calcium carbonate, the same substance that comprises eggshells. Sea-urchin spines are usually hollow and can be fragile, particularly when it comes to extracting broken spines from the skin.

Injuries usually happen when people step on them on while walking across shallow rocky bottoms or tide pools. Divers and snorkelers are often injured while swimming on the surface in shallow waters as well as when entering or exiting the water from shore dives.

Epidemiology

Although little epidemiological data is available, sea-urchin puncture wounds are common among divers, particularly when in shallow waters, near rocky shores or in close proximity to wrecks and other hard surfaces. The DAN Medical Information team receives at least one call a week regarding sea-urchin injuries, typically from divers and snorkelers swimming in very shallow waters near rocky shores.

Signs and Symptoms

Injuries are typically in the form of puncture wounds, often multiple and localized. Skin scrapes and lacerations are also possible. Puncture wounds are generally painful and associated with redness and swelling. Pain ranges from mild to severe depending on several factors, including the species of sea urchin, the body area of the wound, joint or muscular layers compromised, number of punctures, depth of puncture, and the individual's threshold for pain. Multiple puncture wounds may cause limb weakness or paralysis, particularly with the long-spined species of the genus Diadema. On very rare occasions, immediate life-threatening complications may occur.

Prevention

1. Be observant while entering or exiting the water from shore dives, particularly when the bottom is rocky.
2. If swimming, snorkeling or diving in shallow waters, near rocky shores or in close proximity to wrecks and other hard surfaces, maintain a prudent distance and buoyancy control.
3. Avoid handling these animals.

References

- Kaplan, Eugene H.; Kaplan, Susan L.; Peterson, Roger Tory (August 1999). A Field Guide to Coral Reefs: Caribbean and Florida. Boston : Houghton Mifflin. P. 55. ISBN 978-0-618-00211-5
- Mollusk, animal: britannica.com, Retrieved 20 July, 2019
- Crustacean, animal: britannica.com, Retrieved 31 July, 2019
- Dawson, Mike N.; Martin, Laura E.; Penland, Lolita K. (2001). Jellyfish swarms, tourists, and the Christ-child. Hydrobiologia. 451. Pp. 131–144. Doi:10.1023/A:1011868925383. ISBN 978-0-7923-6964-6
- Fish, animal: britannica.com, Retrieved 25 August, 2019
- All-about-aquatic-plants, flower-facts, blog: 1800flowers.com, Retrieved 25 August, 2019
- Ruppert, Edward E.; Fox, Richard, S.; Barnes, Robert D. (2004). Invertebrate Zoology, 7th edition. Cengage Learning. Pp. 148–174. ISBN 978-81-315-0104-7
- Sea-urchins, hazardous-marine-life, health: diversalertnetwork.org, Retrieved 23 June, 2019

Methods and Techniques 4

- Fish Ladder
- Raceway
- Recirculating Aquaculture System
- Aquaponics
- Antimicrobials in Aquaculture
- Copper Alloys in Aquaculture

The numerous techniques used in aquaculture include raceway, aquaponics, recirculating aquaculture systems, etc. All these methods and techniques related to aquaculture as well as practices such as using antimicrobials and copper alloys has been carefully analyzed in this chapter.

Fish Ladder

Fish ladders can be found along important fish migration routes. These structures are located in waterways, usually rivers, alongside manmade or natural obstructions. Dams, waterfalls, and locks are all examples of obstructions that impede fish migration patterns. The fish ladder (also known as fish steps, fish pass, or fishway) assists fish in crossing these difficult obstacles by providing an alternative route. The principal characteristic of fish ladders is in the design, which consists of small, shallow steps that fish are able to jump. These steps progressively climb over the obstacle and allow fish to reach the open waters on the other side. The water running over fish ladders must be controlled so that it attracts the attention of fish, but that it is not too strong for fish to swim against.

Fish ladders can be traced back to the 1600s in France. The first fish ladders were built with bunches of tree limbs that allowed fish to cross difficult channels of water. By 1837, Richard McFarlan patented the first fish ladder. His version was used to help fish cross the dam at his hydro-powered lumber mill in New Brunswick, Canada. The idea traveled across the Atlantic Ocean to Ireland by 1852, when a fish ladder was built to encourage salmon migration where fisheries had been unsuccessful.

Use of fish ladders continued to spread and in 1880, the first fish ladder of the US was built in Rhode Island. It was originally built along the wooden Pawtuxet Falls dam, but was later torn down when the state replaced the wood with concrete in 1924. With the progression of industrialization throughout North America and Europe, the number of locks and dams continued to increase, contributing to a growing need for fish ladders.

Fish Ladder Models

Throughout the years, various fish ladder models have been utilized. These include the following:

- Baffle Fishway: Baffle fishways manipulate the the water current to allow fish to swim around obstructions.
- Rock-ramp Fishway: The rock-ramp fishway uses large rocks and logs to create small pools of water between steps. This allows fish to jump from one pool to the next without fighting a strong current.
- Fish Elevator: A fish elevator begins with a collection area at the bottom of the obstacle. Once the collection area is full, the fish are corralled into a hopper (like an elevator car) that carries them up and over the obstacle.
- Pool and Weir: The pool and weir model is similar to the rock-ramp fishway. Instead of stones and logs, however, this model uses cement to create miniature dams and pools of water. This allows fish to jump from pool to pool and cross the obstacle.
- Fish Siphon: The fish siphon veers off the principal waterway, running parallel to the river. Once it bypasses the obstruction, it rejoins the waterway on the other side.
- Vertical Slot Fish Passage: The vertical slot fish passage is like the pool and weir system. Instead of jumping from pool to pool, however, this design incorporates a slot through which fish can swim to the next step.

Impact on Fish Migration

Analyses of fish ladders have identified varying degrees of effectiveness. Evidence suggests that some fish species are unable to utilize these structures. For example, research shows that only 3% of the American shad are able to cross fish ladders, 97% are unable to reach spawning territories. Additionally, some fish are believed to be affected by the electric currents that pass through metal rods in concrete fish ladders. This is particularly true of the paddlefish which has electric field sensitive organs.

Without fish ladders, however, many more fish would be prevented from reaching their breeding and feeding grounds. As human action often negatively influences wildlife, it is important to have such structures in order to reduce unwanted consequences.

Raceway

A raceway, also known as a flow-through system, is an artificial channel used in aquaculture to culture aquatic organisms. Raceway systems are among the earliest methods used for inland aquaculture. A raceway usually consists of rectangular basins or canals constructed of concrete and equipped with an inlet and outlet. A continuous water flow-through is maintained to provide the required level of water quality, which allows animals to be cultured at higher densities within the raceway.

Freshwater species such as trout, catfish and tilapia are commonly cultured in raceways. Raceways are also used for some marine species which need a constant water flow, such as juvenile salmon, brackish water sea bass and sea bream and marine invertebrates such as abalone.

Site Selection

The most important factor to consider when selecting a site for a raceway farm is the water supply. Water sources for raceway aquaculture operations are usually streams, springs, reservoirs or deep wells. Trout do best in spring water because it keeps a constant temperature, while catfish need a strong flow, about 80 litres per second for every 0.4 hectares of raceway. A backup water supply should be positioned so, if the water supply or pump fails, it can flow by gravity into the start of the raceway.

Construction

Most raceways are made of reinforced concrete, though some earthen raceways are also built. Earthen raceways with plastic liners cost little and are easy to build, but cleaning and disinfecting them is difficult and plastic linings are fragile. Reinforced concrete is more expensive, but is durable and can be shaped in complex ways. Raceway tanks can also be built from polyester resin. These tanks have smooth walls, and are mobile and easy to service. However, their cost limits them to small sizes, under 5 cubic metres.

Size

A raceway is most often a rectangular canal with a water current flowing from a supply end to an exit end. The length to width ratio is important in raceways. To prevent the fish stock from swimming in circular movements, which would cause debris to build up in the centre, a length to width ratio of at least six to one is recommended. If the width is too large this could result in a feeble current speed which is not desirable. The length of a raceway unit is usually constrained by the water quality or by how much stock a unit can hold for ease of management. The average depth of a raceway for fin fish, such as rainbow trout, is about one metre. This means each section in a raceway should be about 30 m long and 2.5–3 m wide. The landscape should sloped to one or two percent, so the flow through the system can be maintained by gravity. The raceway should not be curved, so the flow will be uniform.

A raceway farm for freshwater fin fish usually has a dozen or more parallel raceway strips build alongside each other, with each strip consisting of 15 to 20 or more serial sections. The risk of unhygienic conditions increases towards the lower level sections, and can be kept in check by ensuring there are not too many sections and the water flow is adequate. In order to isolate any diseased section and avoid transmitting the disease back to the upper raceways, each section should have its own drainage channel. Controls, such as weirs, are also needed to ensure individual raceways can't accidentally overflow or empty.

Water Flow

The water flow rate in a raceway system needs to be sufficiently high to meet the respiratory (dissolved oxygen) requirements for the species concerned and to flush out metabolic wastes, especially ammonia. In a well designed system, the existing water in the raceway is largely replaced by new water when the same volume of new water enters the raceway. Self-cleaning can sometimes be achieved if the fish stocks density is sufficiently high and the water level is sufficiently low. For example, if trout are stocked at 20 kilograms per cubic metre, they can keep the raceway unit clean by their swimming movements, preventing waste solids from settling to the raceway floor.

However, in most cases it is necessary to frequently clean raceways. The simplest way is to lower the water level in the raceway units, which increases the speed of the water current, and then herd the fish together till they flush the waste from the raceway. Solid wastes which accumulate at the raceway bottom can be removed by pumps. Oxygen levels in the water can be kept high if the raceway units are placed one after the other with intermediate drops over weirs, or by the use of aeration systems such as pumps, blowers and agitators.

Generally the water should be replaced about every hour. This means a typical raceway section requires a flow rate around 30 liters per second. However, the optimum flow through rate depends on the species, because there are differences in the rates at which oxygen is consumed and metabolic wastes are produced. For example, trout and juvenile salmon are less tolerant of degraded water quality and require a more rapid water turnover than catfish or tilapia. The flow rate necessary to maintain water quality can also change through the year, as the temperature changes and the cultured species grow larger. For reason such as these, continuous monitoring of water quality is important, including measurements of water flow rates, pH levels and temperature, as well as the levels of dissolved oxygen, and suspended and solid waste material.

Maximum Load

The maximum load of organisms that can be cultured in a raceway system depends on the species, and particularly on the size of the species. For trout, stocking rates of 30 to 50 kg/m^3 are normal at the end of a rearing cycle, while for marine species, such as sea bass and sea bream, the achievable load is lower, between 15 and 20 kg/m^3. The total volume required for a raceway is calculated by dividing the total amount of fish in kg by the desired stocking rate in kg per m^3.

Feed

In most raceway aquaculture food needs to be supplied. The composition of the food, and the amount and time of feeding needs to be adjusted to the specific species. This can be optimised to reduce costs and minimize the amount of waste.

Waste Water

The treatment of waste water issuing from raceway farms is a major concern. Fish fecal matter and uneaten feed are typically the major elements of solid waste produced in raceway aquaculture farms. These can adversely impact the environment in the receiving water body. Of particular environmental concern is the waste product phosphorus. Excessive discharge of phosphorus to receiving waters can result in eutrophication. For example, in Korea poor waste treatments in trout farms resulted in reservoirs and rivers developing red tides, which caused wider social problems.

Because raceway aquaculture operations discharge large volumes of water, the concentration of discharged solids is low. This means it is not easy to treat and implement practical, cost effective treatments. Technologies for the removal of solids include microscreens, dual-drain tanks, swirl separators, plate separators, baffles, media filters, air flotation, foam fractionation, chemical flocculation, and constructed wetlands. But because of the impracticality and high costs of these methods, most of them are not applicable for commercial aquaculture. As a consequence, sedimentation (settling) is still the most widely applied and cost effective technology. Since 1999, regulations in South Korea require that all raceway farms provide waste water treatment facilities covering at least 20% of the farmed area to prevent pollution of the freshwater environment. Open race way ponds were used for removal of heavy metal ions like lead using live *Spirulina (Arthospira) sp.*

In Shrimp Farming

Recently Dr. Addison Lawrence received a patent for his System and Method for Super-Intensive Shrimp Production. This system employs artificial raceways to produce large quantities of shrimp. In an interview with Undercurrent News Lawrence said,

"We have several very interested commercial groups interested in using the technology to develop shrimp farms in the United States, and we have several groups interested in developing farms outside of the US," Super-Intensive Shrimp Production offers the capability to have no outflow, saving on water costs, reducing the impact on local water resources and reducing the environmental impact of shrimp farming.

Recirculating Aquaculture System

Recirculating aquaculture systems (RAS) are used in home aquaria and for fish production where water exchange is limited and the use of biofiltration is required to reduce ammonia toxicity. Other types of filtration and environmental control are often also necessary to maintain clean water and provide a suitable habitat for fish. The main benefit of RAS is the ability to reduce the need for fresh, clean water while still maintaining a healthy environment for fish. To be operated economically commercial RAS must have high fish stocking densities, and many researchers are currently conducting studies to determine if RAS is a viable form of intensive aquaculture.

RAS Water Treatment Processes

A biofilter and CO_2 degasser on an outdoor recirculating aquaculture system used to grow largemouth bass

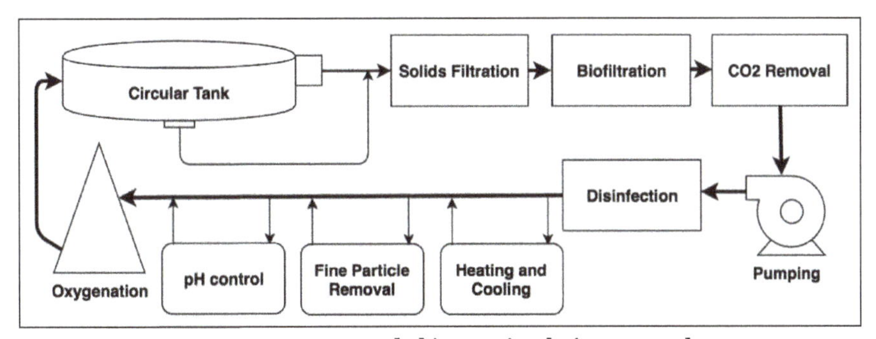

Water treatment processes needed in a recirculating aquaculture system

A series of treatment processes is utilized to maintain water quality in intensive fish farming operations. These steps are often done in order or sometimes in tandem. After leaving the vessel holding fish the water is first treated for solids before entering a biofilter to convert ammonia, next degassing and oxygenation occur, often followed by heating/cooling and sterilization. Each of these processes can be completed by using a variety of different methods and equipment, but regardless all must take place to ensure a healthy environment that maximizes fish growth and health.

Biofiltration

All RAS relies on biofiltration to convert ammonia (NH_4^+ and NH_3) excreted by the fish into nitrate. Ammonia is a waste product of fish metabolism and high concentrations (>.02 mg/L) are toxic to most finfish. Nitrifying bacteria are chemoautotrophs that convert ammonia into nitrite then nitrate. A biofilter provides a substrate for the bacterial community, which results in thick biofilm growing within the filter. Water is pumped through the filter, and ammonia is utilized by the bacteria for energy. Nitrate is less toxic than ammonia (>100 mg/L), and can be removed by a denitrifying biofilter or by water replacement. Stable environmental conditions and regular maintenance are required to ensure the biofilter is operating efficiently.

Solids Removal

In addition to treating the liquid waste excreted by fish the solid waste must also be treated, this is done by concentrating and flushing the solids out of the system. Removing solids reduces bacteria growth, oxygen demand, and the proliferation of disease. The simplest method for removing solids is the creation of settling basin where the relative velocity of the water is slow and particles can settle at the bottom of the tank where they are either flushed out or vacuumed out manually using a siphon. However, this method is not viable for RAS operations where a small footprint is desired. Typical RAS solids removal involves a sand filter or particle filter where solids become lodged and can be periodically backflushed out of the filter. Another common method is the use of a mechanical drum filter where water is run over a rotating drum screen that is periodically cleaned by pressurized spray nozzles, and the resulting slurry is treated or sent down the drain. In order to remove extremely fine particles or colloidal solids a protein fractionator may be used with or without the addition of ozone (O_3).

Oxygenation

Reoxygenating the system water is a crucial part to obtaining high production densities. Fish require oxygen to metabolize food and grow, as do bacteria communities in the biofilter. Dissolved oxygen levels can be increased through two methods aeration and oxygenation. In aeration air is pumped through an air stone or similar device that creates small bubbles in the water column, this results in a high surface area where oxygen can dissolve into the water. In general due to slow gas dissolution rates and the high air pressure needed to create small bubbles this method is considered inefficient and the water is instead oxygenated by pumping in pure oxygen. Various methods are used to ensure that during oxygenation all of the oxygen dissolves into the water column. Careful calculation and consideration must be given to the oxygen demand of a given system, and that demand must be met with either oxygenation or aeration equipment.

pH Control

In all RAS pH must be carefully monitored and controlled. The first step of nitrification in the biofilter consumes alkalinity and lowers the pH of the system. Keeping the pH in a suitable range (5.0-9.0 for freshwater systems) is crucial to maintain the health of both the fish and biofilter. pH is typically controlled by the addition of alkalinity in the form of lime ($CaCO_3$) or sodium hydroxide (NaOH). A low pH will lead to high levels of dissolved carbon dioxide (CO_2), which can prove toxic to fish. pH can also be controlled by degassing CO_2 in a packed column or with an aerator, this is necessary in intensive systems especially where oxygenation instead of aeration is used in tanks to maintain O_2 levels.

Temperature Control

All fish species have a preferred temperature above and below which that fish will experience negative health effects and eventually death. Warm water species such as Tilapia and Barramundi prefer 24 °C water or warmer, where as cold water species such as trout and salmon prefer water temperature below 16 °C. Temperature also plays an important role in dissolved oxygen (DO) concentrations, with higher water temperatures having lower values for DO saturation. Temperature is controlled through the use of submerged heaters, heat pumps, chillers, and heat exchangers. All four may be used to keep a system operating at the optimal temperature for maximizing fish production.

Biosecurity

Disease outbreaks occur more readily when dealing with the high fish stocking densities typically employed in intensive RAS. Outbreaks can be reduced by operating multiple independent systems with the same building and isolating water to water contact between systems by cleaning equipment and personnel that move between systems. Also the use of a Ultra Violet (UV) or ozone water treatment system reduces the number of free floating virus and bacteria in the system water. These treatment systems reduce the disease loading that occurs on stressed fish and thus reduce the chance of an outbreak.

Advantages

- Reduced water requirements as compared to raceway or pond aquaculture systems.
- Reduced land needs due to the high stocking density
- Site selection flexibility and independence from a large, clean water source.
- Reduction in wastewater effluent volume.
- Increased biosecurity and ease in treating disease outbreaks.

- Ability to closely monitor and control environmental conditions to maximize production efficiency. Similarly, independence from weather and variable environmental conditions.

Sturgeon grown at high density in a partial recirculating aquaculture system

Disadvantages

- High upfront investment in materials and infrastructure.
- High operating costs mostly due to electricity, and system maintenance.
- A need for highly trained staff to monitor and operate the system.

Special Types of RAS

Aquaponics

Combining plants and fish in a RAS is referred to as aquaponics. In this type of system ammonia produced by the fish is not only converted to nitrate but is also removed by the plants from the water. In an aquaponics system fish effectively fertilize the plants, this creates a closed looped system where very little waste is generated and inputs are minimized. Aquaponics provides the advantage of being able to harvest and sell multiple crops. Contradictory views exist on the suitability and safety of RAS effluents to sustain plant growth under aquaponics condition. Future conversions, rather 'upgrades', of operational RAS farms to semi-commercial Aquaponic ventures should not be deterred by nutrient insufficiency or nutrient safety arguments. Incentivizing RAS farm wastes through semi-commercial aquaponics is encouraged. Nutrients locked in RAS wastewater and sludge have sufficient and safe nutrients to sustain plant growth under aquaponics condition.

Aquariums

Home aquaria and inland commercial aquariums are a form of RAS where the water quality is very carefully controlled and the stocking density of fish is relatively low. In these systems the goal is to display the fish rather than producing food. However, biofilters and other forms of water treatment are still used to reduce the need to exchange water and to maintain water clarity. Just like in traditional RAS water must be removed periodically to prevent nitrate and other toxic chemicals from building up in the system. Coastal aquariums often have high rates of water exchange and are typically not operated as a RAS due to their proximity to a large body of clean water.

Aquaponics

Aquaponics refers to any system that combines conventional aquaculture (raising aquatic animals such as snails, fish, crayfish or prawns in tanks) with hydroponics (cultivating plants in water) in a symbiotic environment. In normal aquaculture, excretions from the animals being raised can accumulate in the water, increasing toxicity. In an aquaponic system, water from an aquaculture system is fed to a hydroponic system where the by-products are broken down by nitrifying bacteria initially into nitrites and subsequently into nitrates that are utilized by the plants as nutrients. Then, the water is recirculated back to the aquaculture system.

As existing hydroponic and aquaculture farming techniques form the basis for all aquaponic systems, the size, complexity, and types of foods grown in an aquaponic system can vary as much as any system found in either distinct farming discipline.

Aquaponics has ancient roots, although there is some debate on its first occurrence:

- Aztec cultivated agricultural islands known as *chinampas* in a system considered by some to be the first form of aquaponics for agricultural use, where plants were raised on stationary (or sometime movable) islands in lake shallows and waste materials dredged from the Chinampa canals and surrounding cities were used to manually irrigate the plants.
- South China and the whole of Southeast Asia, where rice was cultivated and farmed in paddy fields in combination with fish, are cited as examples of early aquaponics systems, although the technology had been brought by Chinese settlers who had migrated from Yunnan around 5 AD. These polycultural farming systems existed in many Far Eastern countries and raised fish such as the oriental loach, swamp eel, common carp and crucian carp as well as pond snails in the paddies.
- The 13th century Chinese agricultural manual *Wang Zhen's Book on Farming* described floating wooden rafts which were piled with mud and dirt and which were used for growing rice, wild rice, and fodder. Such floating planters were employed in regions constituting the modern provinces of Jiangsu, Zhejiang, and Fujian. These floating planters are known as either *jiatian* or *fengtian*,

which translates to "framed paddy" and "brassica paddy", respectively. The agricultural work also references earlier Chinese texts, which indicated that floating raft rice cultivation was being used as early as the Tang Dynasty (6th century) and Northern Song Dynasty (8th century) periods of Chinese history.

Floating aquaponics systems on polycultural fish ponds have been installed in China in more recent years on a large scale. They are used to grow rice, wheat and canna lily and other crops, with some installations exceeding 2.5 acres (10,000 m^2).

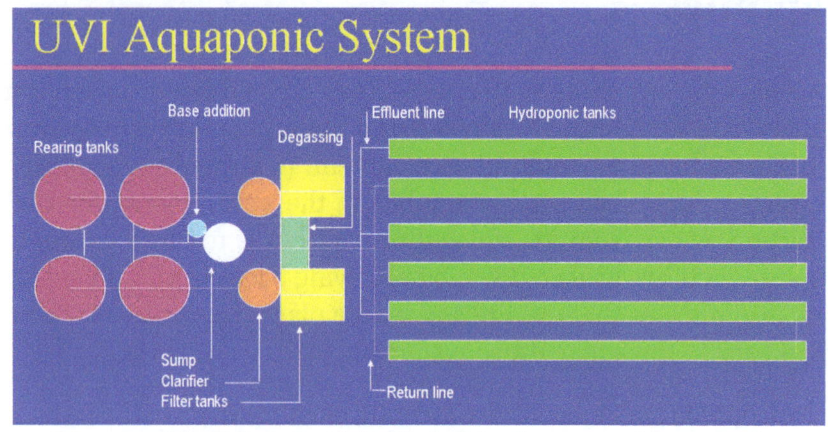

Diagram of the University of the Virgin Islands commercial aquaponics system designed to yield 5 metric tons of Tilapia per year

The development of modern aquaponics is often attributed to the various works of the New Alchemy Institute and the works of Dr. Mark McMurtry et al. at the North Carolina State University. Inspired by the successes of the New Alchemy Institute and the reciprocating aquaponics techniques developed by Dr. Mark McMurtry et al., other institutes soon followed suit. Starting in 1979, Dr. James Rakocy and his colleagues at the University of the Virgin Islands researched and developed the use of deep water culture hydroponic grow beds in a large-scale aquaponics system.

The first aquaponics research in Canada was a small system added onto existing aquaculture research at a research station in Lethbridge, Alberta. Canada saw a rise in aquaponics setups throughout the '90s, predominantly as commercial installations raising high-value crops such as trout and lettuce. A setup based on the deep water system developed at the University of Virgin Islands was built in a greenhouse at Brooks, Alberta where Dr. Nick Savidov and colleagues researched aquaponics from a background of plant science. The team made findings on rapid root growth in aquaponics systems and on closing the solid-waste loop, and found that owing to certain advantages in the system over traditional aquaculture, the system can run well at a low pH level, which is favoured by plants but not fish.

Parts of an Aquaponic System

A commercial aquaponics system. An electric pump moves nutrient-rich water from the fish tank through a solids filter to remove particles the plants above cannot absorb. The water then provides nutrients for the plants and is cleansed before returning to the fish tank below.

Aquaponics consists of two main parts, with the aquaculture part for raising aquatic animals and the hydroponics part for growing plants. Aquatic effluents, resulting from uneaten feed or raising animals like fish, accumulate in water due to the closed-system recirculation of most aquaculture systems. The effluent-rich water becomes toxic to the aquatic animal in high concentrations but this contains nutrients essential for plant growth. Although consisting primarily of these two parts, aquaponics systems are usually grouped into several components or subsystems responsible for the effective removal of solid wastes, for adding bases to neutralize acids, or for maintaining water oxygenation. Typical components include:

- Rearing tank: The tanks for raising and feeding the fish;
- Settling basin: A unit for catching uneaten food and detached biofilms, and for settling out fine particulates;
- Biofilter: A place where the nitrification bacteria can grow and convert ammonia into nitrates, which are usable by the plants;
- Hydroponics subsystem: The portion of the system where plants are grown by absorbing excess nutrients from the water;
- Sump: The lowest point in the system where the water flows to and from which it is pumped back to the rearing tanks.

Depending on the sophistication and cost of the aquaponics system, the units for solids removal, biofiltration, and the hydroponics subsystem may be combined into one unit or subsystem, which prevents the water from flowing directly from the aquaculture part of the system to the hydroponics part. By utilizing gravel or sand as plant supporting medium, solids are captured and the medium has enough surface area for fixed-film nitrification. The ability to combine biofiltration and hydroponics allows for aquaponic system to in many cases eliminate the need for an expensive, separate biofilter.

Live Components

An aquaponic system depends on different live components to work successfully. The three main live components are plants, fish (or other aquatic creatures) and bacteria. Some systems also include additional live components like worms.

Plants

Many plants are suitable for aquaponic systems, though which ones work for a specific system depends on the maturity and stocking density of the fish. These factors influence the concentration of nutrients from the fish effluent and how much of those nutrients are made available to the plant roots via bacteria. Green leaf vegetables with low to medium nutrient requirements are well adapted to aquaponic systems, including chinese cabbage, lettuce, basil, spinach, chives, herbs, and watercress.

A Deep Water Culture hydroponics system where plant grow directly into the effluent rich water without a soil medium. Plants can be spaced closer together because the roots do not need to expand outwards to support the weight of the plant

Other plants, such as tomatoes, cucumbers, and peppers, have higher nutrient requirements and will do well only in mature aquaponic systems with high stocking densities of fish.

Plants that are common in salads have some of the greatest success in aquaponics, including cucumbers, shallots, tomatoes, lettuce, chiles, capsicum, red salad onions and snow peas. Some profitable plants for aquaponic systems include chinese cabbage, lettuce, basil, roses, tomatoes, okra, cantaloupe and bell peppers.

Plant placed into a nutrient rich water channel in a Nutrient film technique (NFT) system

Spinach seedlings, 5 days old, by aquaponics

Other species of vegetables that grow well in an aquaponic system include watercress, basil, coriander, parsley, lemongrass, sage, beans, peas, kohlrabi, taro, radishes, strawberries, melons, onions, turnips, parsnips, sweet potato, cauliflower, cabbage, broccoli, and eggplant as well as the choys that are used for stir fries.

Fish or other Aquatic Creatures

Filtered water from the hydroponics system drains into a catfish tank for re-circulation

Freshwater fish are the most common aquatic animal raised using aquaponics due to their ability to tolerate crowding, although freshwater crayfish and prawns are also sometimes used. There is a branch of aquaponics using saltwater fish, called saltwater aquaponics. There are many species of warmwater and coldwater fish that adapt well to aquaculture systems.

In practice, tilapia are the most popular fish for home and commercial projects that are intended to raise edible fish because it is a warmwater fish species that can tolerate crowding and changing water conditions. Barramundi, silver perch, eel-tailed catfish or tandanus catfish, jade perch and Murray cod are also used. For temperate climates when there isn't ability or desire to maintain water temperature, bluegill and catfish are suitable fish species for home systems.

Koi and goldfish may also be used, if the fish in the system need not be edible. Other suitable fish include channel catfish, rainbow trout, perch, common carp, Arctic char, largemouth bass and striped bass.

Bacteria

Nitrification, the aerobic conversion of ammonia into nitrates, is one of the most important functions in an aquaponic system as it reduces the toxicity of the water for fish, and allows the resulting nitrate compounds to be removed by the plants for nourishment. Ammonia is steadily released into the water through the excreta and gills of fish as a product of their metabolism, but must be filtered out of the water since higher concentrations of ammonia (commonly between 0.5 and 1 ppm) can impair growth, cause widespread damage to tissues, decrease resistance to disease and even kill the fish. Although plants can absorb ammonia from the water to some degree, nitrates are assimilated more easily, thereby efficiently reducing the toxicity of the water for fish. Ammonia can be converted into safer nitrogenous compounds through combined healthy populations of 2 types of bacteria: *Nitrosomonas* which convert ammonia into nitrites, and *Nitrobacter* which then convert nitrites into nitrates. While nitrate is still harmful to fish due to its ability to create metehemoglobine, which cannot bind oxygen, by attaching to hemoglobin, nitrates are able to be tolerated at high levels by fish. High surface area provides more space for the growth of nitrifying bacteria. Grow bed material choices require careful analysis of the surface area, price and maintainability considerations.

Hydroponic Subsystem

Plants are grown as in hydroponics systems, with their roots immersed in the nutrient-rich effluent water. This enables them to filter out the ammonia that is toxic to the aquatic animals, or its metabolites. After the water has passed through the hydroponic subsystem, it is cleaned and oxygenated, and can return to the aquaculture vessels. This cycle is continuous. Common aquaponic applications of hydroponic systems include:

- Deep-water raft aquaponics: Styrofoam rafts floating in a relatively deep aquaculture basin in troughs. Raft tanks can be constructed to be quite large, and enable seedlings to be transplanted at one end of the tank while fully grown plants are harvested at the other, thus ensuring optimal floor space usage.
- Recirculating aquaponics: Solid media such as gravel or clay beads, held in a container that is flooded with water from the aquaculture. This type of aquaponics is also known as closed-loop aquaponics.
- Reciprocating aquaponics: Solid media in a container that is alternately flooded

and drained utilizing different types of siphon drains. This type of aquaponics is also known as flood-and-drain aquaponics or ebb-and-flow aquaponics.

- Nutrient film technique channels: Plants are grown in lengthy narrow channels, with a film of nutrient-filled water constantly flowing past the plant roots. Due to the small amount of water and narrow channels, helpful bacteria cannot live there and therefore a bio filter is required for this method.
- Other systems use towers that are trickle-fed from the top, horizontal PVC pipes with holes for the pots, plastic barrels cut in half with gravel or rafts in them. Each approach has its own benefits.

Since plants at different growth stages require different amounts of minerals and nutrients, plant harvesting is staggered with seedlings growing at the same time as mature plants. This ensures stable nutrient content in the water because of continuous symbiotic cleansing of toxins from the water.

Biofilter

In an aquaponics system, the bacteria responsible for the conversion of ammonia to usable nitrates for plants form a biofilm on all solid surfaces throughout the system that are in constant contact with the water. The submerged roots of the vegetables combined have a large surface area where many bacteria can accumulate. Together with the concentrations of ammonia and nitrites in the water, the surface area determines the speed with which nitrification takes place. Care for these bacterial colonies is important as to regulate the full assimilation of ammonia and nitrite. This is why most aquaponics systems include a biofiltering unit, which helps facilitate growth of these microorganisms. Typically, after a system has stabilized ammonia levels range from 0.25 to .50 ppm; nitrite levels range from 0.0 to 0.25 ppm, and nitrate levels range from 5 to 150 ppm. During system startup, spikes may occur in the levels of ammonia (up to 6.0 ppm) and nitrite (up to 15 ppm), with nitrate levels peaking later in the startup phase. In the nitrification process ammonia is oxidized into nitrite, which releases hydrogen ions into the water. Overtime your pH will slowly drop, so you can use non-sodium bases such as potassium hydroxide or calcium hydroxide to neutralize the water's pH if insufficient quantities are naturally present in the water to provide a buffer against acidification. In addition, selected minerals or nutrients such as iron can be added in addition to the fish waste that serves as the main source of nutrients to plants.

A good way to deal with solids buildup in aquaponics is the use of worms, which liquefy the solid organic matter so that it can be utilized by the plants and other animals in the system.

Operation

The five main inputs to the system are water, oxygen, light, feed given to the aquatic animals, and electricity to pump, filter, and oxygenate the water. Spawn or fry may be added to replace grown fish that are taken out from the system to retain a stable system. In terms of outputs, an aquaponics system may continually yield plants such as vegetables grown in hydroponics, and edible aquatic species raised in an aquaculture. Typical build ratios are .5 to 1 square foot of grow space for every 1 U.S. gal (3.8 L) of aquaculture water in the system. 1 U.S. gal (3.8 L) of water can support between .5 lb (0.23 kg) and 1 lb (0.45 kg) of fish stock depending on aeration and filtration.

Ten primary guiding principles for creating successful aquaponics systems were issued by Dr. James Rakocy, the director of the aquaponics research team at the University of the Virgin Islands, based on extensive research done as part of the *Agricultural Experiment Station* aquaculture program:

- Use a feeding rate ratio for design calculations.
- Keep feed input relatively constant.
- Supplement with calcium, potassium and iron.
- Ensure good aeration.
- Remove solids.
- Be careful with aggregates.
- Oversize pipes.
- Use biological pest control.
- Ensure adequate biofiltration.
- Control pH.

Feed Source

As in most aquaculture based systems, stock feed often consists of fish meal derived from lower-value species. Ongoing depletion of wild fish stocks makes this practice unsustainable. Organic fish feeds may prove to be a viable alternative that relieves this concern. Other alternatives include growing duckweed with an aquaponics system that feeds the same fish grown on the system, excess worms grown from vermiculture composting, using prepared kitchen scraps, as well as growing black soldier fly larvae to feed to the fish using composting grub growers.

Plant Nutrients

Like hydroponics, a few minerals and micro nutrients can be added to improve plants growth. Iron is the most deficient nutrient in aquaponics, it can be added through mixing

Iron Chelate powder with water. Potassium can be added as potassium sulfate through foliar spray. Less vital nutrients include epsom salt, calcium chloride and boron. Biological filtration of aquaculture wastes yield high nitrate concentrations, which is great for leafy greens. For flowering plants with high nutrient demands it is recommended to introduce supplemental nutrients such as magnesium, calcium, potassium, and phosphorus. Common sources are sulfate of potash, potassium bicarbonate, monoammonium phosphate, etc. Nutrient deficiency in wastewater from fish component (RAS) can be completely masked using raw or mineralized sludge, usually containing 3–17 times higher nutrient concentrations. RAS effluents (wastewater and sludge combined) contain adequate N, P, Mg, Ca, S, Fe, Zn, Cu, Ni to meet most aquaponic crop needs. Potassium is generally deficient requiring full-fledged fertilization. Micronutrients B, Mo are partly sufficient and can be easily ameliorated by increasing sludge release. The presumption surrounding 'definite' phyto-toxic sodium levels in RAS effluents should be reconsidered – practical solutions available too. No threat of heavy metal accumulation exists within the aquaponics loop.

Water Usage

Aquaponic systems do not typically discharge or exchange water under normal operation, but instead recirculate and reuse water very effectively. The system relies on the relationship between the animals and the plants to maintain a stable aquatic environment that experience a minimum of fluctuation in ambient nutrient and oxygen levels. Plants are able to recover dissolved nutrients from the circulating water, meaning that less water is discharged and the water exchange rate can be minimized. Water is added only to replace water loss from absorption and transpiration by plants, evaporation into the air from surface water, overflow from the system from rainfall, and removal of biomass such as settled solid wastes from the system. As a result, aquaponics uses approximately 2% of the water that a conventionally irrigated farm requires for the same vegetable production. This allows for aquaponic production of both crops and fish in areas where water or fertile land is scarce. Aquaponic systems can also be used to replicate controlled wetland conditions. Constructed wetlands can be useful for biofiltration and treatment of typical household sewage. The nutrient-filled overflow water can be accumulated in catchment tanks, and reused to accelerate growth of crops planted in soil, or it may be pumped back into the aquaponic system to top up the water level.

Energy Usage

Aquaponic installations rely in varying degrees on man-made energy, technological solutions, and environmental control to achieve recirculation and water/ambient temperatures. However, if a system is designed with energy conservation in mind, using alternative energy and a reduced number of pumps by letting the water flow downwards as much as possible, it can be highly energy efficient. While careful design can minimize the risk, aquaponics systems can have multiple 'single points of failure' where problems such as an electrical failure or a pipe blockage can lead to a complete loss of fish stock.

Fish Stocking

In order for aquaponic systems to be financially successful and make a profit whilst also covering its operating expenses, the hydroponic plant components and fish rearing components need to almost constantly be at maximum production capacity. To keep the bio-mass of fish in the system at its maximum (without limiting fish growth), there are 3 main stocking method that can help maintain this maximum:

- Sequential rearing: Multiple age groups of fish share a rearing tank, and when an age group reaches market size they are selectively harvested and replaced with the same amount of fingerlings. Downsides to this method include stressing out the entire pool of fish during each harvest, missing fish resulting in a waste of food/space, and the difficulty of keeping accurate records with frequent harvests.
- Stock splitting: Large quantities of fingerlings are stocked at once and then split into two groups once the tank hits maximum capacity, which is easier to record and eliminates fish being "forgotten". A stress-free way of doing this operation is via "swimways" that connect various rearing tanks and a series of hatches/ moving screens/pumps that move the fish around.
- Multiple rearing units: Entire groups of fish are moved to larger rearing tanks once their current tank hits maximum capacity. Such systems usually have 2–4 tanks that share a filtration system, and when the largest tank is harvested, the other fish groups are each moved up into a bigger tank whilst the smallest tank is restocked with fingerlings. It is also common for there to be several rearing tanks yet no ways to move fish between them, which eliminates the labor of moving fish and allows each tank to be undisturbed during harvesting, even if the space usage is inefficient when the fish are fingerlings.

Ideally the bio-mass of fish in the rearing tanks doesn't exceed 0.5 lbs/gallon, in order to reduce stress from crowding, efficiently feed the fish, and promote healthy growth.

Disease and Pest Management

Although pesticides can normally be used to take care of insect on crops, in an aquaponic system the use of pesticides would threaten the fish ecosystem. On the other hand, if the fish acquire parasites or diseases, therapeutants cannot be used as the plants would absorb them. In order to maintain the symbiotic relationship between the plants and the fish, non-chemical methods such as traps, physical barriers and biological control (such as parasitic wasps/ladybugs to control white flies/aphids) should be used to control pests. The most effective organic pesticide is Neem oil, but only in small quantity to minimize spill over fish's water.

Automation, Monitoring and Control

Many have tried to create automatic control and monitoring systems and some of these demonstrated a level of success. For instance, researchers were able to introduce automation in a small scale aquaponic system to achieve a cost-effective and sustainable farming system. Commercial development of automation technologies has also emerged. For instance, a company has developed a system capable of automating the repetitive tasks of farming and features a machine learning algorithm that can automatically detect and eliminate diseased or underdeveloped plants. A 3.75-acre aquaponics facility that claims to be the first indoor salmon farm in the United States also includes an automated technology. The aquaponic machine has made notable strides in the documenting and gathering of information regarding aquaponics.

Economic Viability

Aquaponics offers a diverse and stable polyculture system that allows farmers to grow vegetables and raise fish at the same time. By having two sources of profit, farmers can continue to earn money even if the market for either fish or plants goes through a low cycle. The flexibility of an aquaponic system allows it to grow a large variety of crops including ordinary vegetables, herbs, flowers and aquatic plants to cater to a broad spectrum of consumers. Herbs, lettuce and speciality greens such as basil or spinach are especially well suited for aquaponic systems due to their low nutritional needs. For the growing number of environmentally conscious consumers, products from aquaponic systems are organic and pesticide free, whilst also leaving a small environmental footprint. Aquaponic systems additionally are economically efficient due to low water usage, effective nutrient cycling and needing little land to operate. Because soil isn't needed and only a little bit of water is required, aquaponic systems can be set up in areas that have traditionally poor soil quality or contaminated water. More importantly, aquaponic systems are usually free of weeds, pests and diseases that would affect soil, which allows them to consistently and quickly produce high quality crops to sell.

Current Examples

The Caribbean island of Barbados created an initiative to start aquaponics systems at home, called the aquaponic machine, with revenue generated by selling produce to tourists in an effort to reduce growing dependence on imported food.

Dakota College at Bottineau in Bottineau, North Dakota has an aquaponics program that gives students the ability to obtain a certificate or an AAS degree in aquaponics.

In Bangladesh, the world's most densely populated country, most farmers use agrochemicals to enhance food production and storage life, though the country lacks oversight on safe levels of chemicals in foods for human consumption. To combat this issue, a team led by M.A. Salam at the Department of Aquaculture of Bangladesh Agricultural University has created plans for a low-cost aquaponics system to provide organic produce and fish for people living in adverse climatic conditions such as the salinity-prone southern area and the flood-prone haor area in the eastern region. Salam's work innovates a form of subsistence farming for micro-production goals at the community and personal levels whereas design work by Chowdhury and Graff was aimed exclusively at the commercial level, the latter of the two approaches take advantage of economies of scale.

Vegetable production part of the low-cost Backyard Aquaponics System developed at Bangladesh Agricultural University

With more than a third of Palestinian agricultural lands in the Gaza Strip turned into a buffer zone by Israel, an aquaponic gardening system is developed appropriate for use on rooftops in Gaza City.

The Smith Road facility in Denver started pilot program of aquaponics to feed 800 to 1000 inmates at Denver Jail and neighboring downtown facility which consist of 1,500 inmates and 700 officers.

In Malaysia Alor Gajah, Melaka, Organization 'Persatuan Akuakutur Malaysia' takes innovative approach in aquaponics by growing Lobster in aquaponics. VertiFarms in New Orleans targets corporate rooftops for vertical farming, accruing up to 90 corporate clients for rooftop vertical farming in 2013.

Windy Drumlins Farm in Wisconsin redesigns aquaponic-solar greenhouse for extreme weather conditions which can endure extremely cold climate. Volunteer operation in Nicaragua "Amigos for Christ" manages its plantation for feeding 900+ poverty-stricken school children by using nutrients from aquaponics method.

Aquaponics in India aims to provide aspiring farmers with aquaponics solutions for commercial and backyard operation. Verticulture in Bedstuy utilizes old Pfizer manufacturing plant for producing basil in commercial scale through aquaponics, yielding 30-40 pounds of basil a week.

Aquaponics startup Edenworks in New York expands to full-scale commercial facility, which will generate 130,000 pounds of greens and 50,000 pounds of fish a year.

There has been a shift towards community integration of aquaponics, such as the non-profit foundation Growing Power that offers Milwaukee youth job opportunities and training while growing food for their community. The model has spawned several satellite projects in other cities, such as New Orleans where the Vietnamese fisherman community has suffered from the Deepwater Horizon oil spill, and in the South Bronx in New York City.

Whispering Roots is a non-profit organization in Omaha, Nebraska that provides fresh, locally grown, healthy food for socially and economically disadvantaged communities by using aquaponics, hydroponics and urban farming.

In addition, aquaponic gardeners from all around the world are gathering in online community sites and forums to share their experiences and promote the development of this form of gardening as well as creating extensive resources on how to build home systems.

Recently, aquaponics has been moving towards indoor production systems. In cities like Chicago, entrepreneurs are utilizing vertical designs to grow food year round. These systems can be used to grow food year round with minimal to no waste.

There are various modular systems made for the public that utilize aquaponic systems to produce organic vegetables and herbs, and provide indoor decor at the same time. These systems can serve as a source of herbs and vegetables indoors. Universities are promoting research on these modular systems as they get more popular among city dwellers.

Antimicrobials in Aquaculture

Antimicrobials destroy bacteria, viruses, fungi, algae, and other microbes. The cells of bacteria (prokaryotes), such as salmonella, differ from those of higher-level organisms (eukaryotes), such as fish. Antibiotics are chemicals designed to either kill or inhibit the growth of pathogenic bacteria while exploiting the differences between prokaryotes and eukaryotes in order to make them relatively harmless in higher-level organisms. Antibiotics are constructed to act in one of three ways: by disrupting cell membranes of bacteria (rendering them unable to regulate themselves), by impeding DNA or protein synthesis, or by hampering the activity of certain enzymes unique to bacteria.

Antibiotics are used in aquaculture to treat diseases caused by bacteria. Sometimes the antibiotics are used to treat diseases, but more commonly antibiotics are used to prevent diseases by treating the water or fish before disease occurs. While this prophylactic method of preventing disease is profitable because it prevents loss and allows fish to grow more quickly, there are several downsides.

The overuse of antibiotics can create antibiotic-resistant bacteria. Antibiotic-resistant bacteria can spontaneously arise when selective pressure to survive results in changes to the DNA sequence of a bacterium allowing that bacterium to survive antibiotic treatments. Because some of the same antibiotics are used to treat fish that are used to treat human disease, pathogenic bacteria causing human disease can also become resistant to antibiotics as a result of treatment of fish with antibiotics. For this reason, the overuse of antibiotics in treatment of fish aquaculture (among other agricultural uses) could create public health issues.

The issue has two sides. In some foreign countries, clean water supplies for aquaculture are extremely limited. Untreated animal manure and human waste are used as feed in shrimp farms and tilapia farms in China and Thailand, in addition to the collection of waste products accumulating from inadequate sewage treatment. In order to prevent the spread of bacteria and disease in contaminated water, some foreign fish farms put U.S.-banned antibiotics into their fishmeal. However, because the more stringent growing regulations in the US increase the price of food, imports from nations without these regulations are increasing based on price and profit.

Between 1995 and 2005, the first ten years of the NAFTA-WTO era in the US, seafood imports increased 65 percent and shrimp imports increased 95 percent. Today, 80 percent of American seafood is imported, about half coming from aquaculture. China, Thailand and Vietnam together account for 44 percent of seafood imports into the United States.

The FDA has been testing for chemicals in aquaculture products for over two decades. In November 2005, the testing program for aquaculture drugs was revised to include antibiotics such as chloramphenicol, fluoroquinolones, nitrofurans, and quinolones, as well as antimicrobial compounds like malachite green that are not approved for use in aquaculture fish. FDA tested samples of catfish, basa, shrimp, dace, and eel from China, finding twenty-five percent of the samples to contain drug residues. FDA has approved five different drugs for use in aquaculture as long as the seafood contains less than a mandated maximum residue limit: florfenicol, sulfamerazine, chorionic gonadotropin, oxytetracycline dihydrate, oxytetracycline hydrochloride, as well as a drug combination of sulfadimethoxine and ormetoprim. FDA has approved two drugs—formalin and hydrogen peroxide—for which it has not set a tolerance.

The FDA now enforces regulations in the US requiring testing of certain imported products for antimicrobial agents under Import Alert 16-131. The Import Alert provides that the use of antimicrobials during the various stages of aquaculture, including malachite green, nitrofurans, fluoroquinolones, and gentian violet, may contribute to an increase of antimicrobial resistance in human pathogens and that prolonged exposure to nitrofurans, malachite green, and gentian violet has been shown to have a carcinogenic affect. In a consumer brochure, the FDA describes the reasoning for enforcement under the import alert:

- After FDA repeatedly found that farm-raised seafood from China was contaminated, the agency announced on June 28, 2007, a broader import control of all farm-raised catfish, basa, shrimp, dace(related to carp), and eel from China. During targeted sampling, from October 2006 through May 2007, FDA repeatedly found that farm-raised seafood from China was contaminated with antimicrobial agents that are not approved for use in the United States. More specifically, the antimicrobials nitrofuran, malachite green, gentian violet, and fluoroquinolones, were detected.
- Due to limitations on funding and resources, U.S. Government Accountability Office states that only 1% of seafood, compared with 2% of all imports, is inspected and only 0.1% of all seafood is tested for antibiotic residue.

Example Antimicrobials

Copper Alloys

Recently, copper alloys have become important netting materials in aquaculture (the farming of aquatic organisms including fish farming). Various other materials including nylon, polyester, polypropylene, polyethylene, plastic-coated welded wire, rubber, patented twine products (Spectra, Dyneema), and galvanized steel are also used for netting in aquaculture fish enclosures around the world. All of these materials are selected for a variety of reasons, including design feasibility, material strength, cost, and corrosion resistance.

A copper alloy pen that has been deployed on a fish farm at depth of 14 feet for one year shows no signs of biofouling

What sets copper alloys apart from the other materials used in fish farming is that copper alloys are antimicrobial. In the marine environment, the antimicrobial/algaecidal properties of copper alloys prevent biofouling, which can briefly be described as the undesirable accumulation, adhesion, and growth of microorganisms, plants, algae, tube worms, barnacles, mollusks, and other organisms on man-made marine structures. By inhibiting microbial growth, copper alloy aquaculture pens avoid the need for costly net changes that are necessary with other materials. The resistance of organism growth on copper alloy nets also provides a cleaner and healthier environment for farmed fish to grow and thrive. In addition to their antifouling benefits, copper alloys have strong structural and corrosion-resistant properties in marine environments. Brass alloy netting cages are also currently being deployed in commercial-scale aquaculture operations in Asia, South America, and the USA.

Methylene Blue

Methylene blue is used in aquaculture and by tropical fish hobbyists as a treatment for fungal infections. It can also be effective in treating fish infected with *ich*, the parasitic protozoa *Ichthyophthirius multifiliis*. It is usually used to protect newly laid fish eggs from being infected by fungus or bacteria. This is useful when the hobbyist wants to artificially hatch the fish eggs. Methylene Blue is also very effective when used as part of a "medicated fish bath" for treatment of ammonia, nitrite, and cyanide poisoning as well as for topical and internal treatment of injured or sick fish as a "first response".

Ozone

Ozone is added to seawater and used for the surface disinfection of haddock and Atlantic halibut eggs against nodavirus. Nodavirus is a lethal and vertically transmitted virus which causes severe mortality in fish. Haddock eggs should not be treated with high ozone level as eggs so treated did not hatch and died after 3–4 days.

Some Problematic Antimicrobials

Malachite Green

In 1983, the FDA banned the use of malachite green in aquaculture. Toxicity studies have shown that this chemical can have serious toxic side effects. Malachite green is not actually an antibiotic, but has antibiotic properties. Malachite green is somewhat stable within the environment and, therefore, is detectable in fish that were treated with the chemical at some point even after treatment has discontinued. After more stringent testing and inspection by the governments of Western Countries, the use of malachite green began to wane and other drugs began to become more prevalent.

Chloramphenicol

While the U.S. has tested farm-raised shrimp for chloramphenicol since 1994, over the last decade the FDA developed a more sensitive testing methodology and changed the levels of detection for chloramphenicol in response to increasing discovery of traces of chloramphenicol in imports. In response to the US discovery of chloramphenicol in imported shrimp and subsequent increased testing sensitivity, the use of this compound in aquaculture began to decrease.

Gentian Violet

Gentian violet, also known as crystal violet has antibacterial, antifungal, and antiparasitic properties. This compound was used during the World War I era as a topical antiseptic, but has been replaced in modern times with more modern treatments. The FDA prohibits the use of gentian violet in aquaculture because of numerous studies showing increased risk of certain cancers related to the compound and a showing that the chemical is bioavailable in fish when used in aquaculture.

Nitrofurans

Nitrofurans are broad spectrum antibiotics, being effective against Gram-positive and Gram-negative bacteria. In 1991, the FDA withdrew several approved food animal nitrofuran products as a result of research showing nitrofurazone, one of the nitrofurans, can produce mammary tumors in rats and ovarian tumors in mice. The FDA also concluded that some people may be hypersensitive to this product. The FDA states, "Absolutely, no extra-label use of the nitrofurans is permitted in any food animals, including seafood." The FDA currently detains certain seafood imports without physical examination due to nitrofuran use by the producer.

Floroquinolones

Fluoroquinolones have been prohibited from extra-label use in the U.S. and many other parts of the world in aquaculture because of public health concern about the development of such antimicrobial resistance. Chinese authorities have acknowledged permitting the use of fluoroquinolones in aquaculture, even though the use of fluoroquinolones in food animals may increase antibiotic resistance in human pathogens compromising the effectiveness of the use of this critically important class of antibiotics in human medicine. The Chinese government has established a higher maximum residue limit than the US and research in China has shown that the Chinese are effectively meeting the Chinese limits. Because of concerns about the presence of fluoroquinolones in the food supply, not only in aquaculture, but also in foods like honey, the U.S. is continuing to develop methods and strategies to detect illegal residues and prevent their introduction into the U.S. food supply.

Copper Alloys in Aquaculture

Copper alloys are important netting materials in aquaculture (the farming of aquatic organisms including fish farming). Various other materials including nylon, polyester, polypropylene, polyethylene, plastic-coated welded wire, rubber, patented twine products (Spectra, Dyneema), and galvanized steel are also used for netting in aquaculture fish enclosures around the world. All of these materials are selected for a variety of reasons, including design feasibility, material strength, cost, and corrosion resistance.

What sets copper alloys apart from the other materials used in fish farming is that copper alloys are antimicrobial, that is, they destroy bacteria, viruses, fungi, algae, and other microbes.

In the marine environment, the antimicrobial/algaecidal properties of copper alloys prevent biofouling, which can briefly be described as the undesirable accumulation, adhesion, and growth of microorganisms, plants, algae, tube worms, barnacles, mollusks, and other organisms on man-made marine structures. By inhibiting microbial growth, copper alloy aquaculture pens avoid the need for costly net changes that are necessary with other materials. The resistance of organism growth on copper alloy nets also provides a cleaner and healthier environment for farmed fish to grow and thrive.

In addition to their antifouling benefits, copper alloys have strong structural and corrosion-resistant properties in marine environments.

It is the combination of all of these properties – antifouling, high strength, and corrosion resistance – that has made copper alloys a desirable material for such marine applications as condenser tubing, water intake screens, ship hulls, offshore structure, and sheathing. In the past 25 years or so, the benefits of copper alloys have caught the attention of the marine aquaculture industry. The industry is now actively deploying copper alloy netting and structural materials in commercial large-scale fish farming operations around the world.

Importance of Aquaculture

Much has been written about the degradation and depletion of natural fish stocks in rivers, estuaries, and the oceans. Because industrial fishing has become extremely efficient, ocean stocks of large fish, such as tuna, cod, and halibut have declined by 90% in the past 50 years.

Aquaculture, an industry that has emerged only in recent decades, has become one of the fastest growing sectors of the world food economy. Aquaculture already supplies more than half of the world's demand for fish. This percentage is predicted to increase dramatically over the next few decades.

Problem of Biofouling

Biofouling is one of the biggest problems in aquaculture. Biofouling occurs on non-copper materials in the marine environment, including fish pen surfaces and nettings. For example, it was noted that the open area of a mesh immersed for only seven days in a Tasmanian aquaculture operation decreased by 37% as a result of biofouling.

Copper alloy mesh installed at an Atlantic salmon fish farm in Tasmania. Foreground: the chain link copper alloy mesh resting on a dock. Distant background: copper alloy mesh pens are installed on the fish farm

The biofouling process begins when algae spores, marine invertebrate larvae, and other organic material adhere to surfaces submerged in marine environments (e.g., fish nets in aquaculture). Bacteria then encourage the attachment of secondary unwanted colonizers.

Biofouling has strong negative impacts on aquaculture operations. Water flow and dissolved oxygen are inhibited due to clogged nets in fish pens. The end result is often diseased fish from infections, such as netpen liver disease, amoebic gill disease, and parasites. Other negative impacts include increased fish mortalities, decreased fish growth rates, premature fish harvesting, reduced fish product values and profitability, and an adversely impacted environment near fish farms.

Biofouling adds enormous weight to submerged fish netting. Two hundredfold increases in weight have been reported. This translates, for example, to two thousand pounds of unwanted organisms adhered to what was once a clean 10-pound fish pen net. In South Australia, biofouling weighing 6.5 tonnes (approximately 13,000 pounds) was observed on a fish pen net. This extra burden often results in net breakage and additional maintenance costs.

To combat parasites from biofouling in finfish aquaculture, treatment protocols such as cypermethrin, azamethiphos, and emamectin benzoate may be administered, but these have been found to have detrimental environmental effects, for example, in lobster operations.

To treat diseases in fish raised in biofouled nets, fish stocks are administered antibiotics. The antibiotics can have unwanted long-term health effects on consumers and on coastal environments near aquaculture operations. To combat biofouling, operators often implement costly maintenance measures, such as frequent net changing, cleaning/removal of unwanted organisms from nets, net repairs, and chemical treatment including antimicrobial coatings on nylon nets. The cost of antifouling a single salmon net can be several thousand British pounds. In some sectors of the European aquaculture industry, cleaning biofouled fish and shellfish pens can cost 5–20% of its market value. Heavy fouling can reduce the saleable product in nets by 60–90%.

Antifouling coatings are often used on nylon nets because the process is more economical than manual cleaning. When nylon nets are coated with antifouling compounds, the coatings repel biofouling for a period of time, usually between several weeks to several months. However, the nets eventually succumb to biofouling. Antifouling coatings containing cuprous oxide algaecide/biocide are the coatings technology used almost exclusively in the fish farming industry today. The treatments usually flake off within a few weeks to six to eight months.

Biofouled nets are replaced after several months of service, depending on environmental conditions, in a complicated, costly, and labor-intensive operation that involves divers and specialized personnel. During this process, live fish in nets must be transferred to clean pens, which causes undue stress and asphyxiation that results in some loss of fish. Biofouled nets that can be reused are washed on land via manual brushing and scrubbing or high-pressure water hosing. They are then dried and re-impregnated with antifouling coatings.

A line of net cleaners is available for in-situ washings where permitted. But, even where not permitted by environmental, fisheries, maritime, and sanitary authorities, should the lack of dissolved oxygen in submerged pens create an emergency condition that endangers the health of fish, divers may be deployed with special in situ cleaning machinery to scrub biofouled nets.

The aquaculture industry is addressing the negative environmental impacts from its operations. As the industry evolves, a cleaner, more sustainable aquaculture industry is expected to emerge, one that may increasingly rely on materials with anti-fouling, anti-corrosive, and strong structural properties, such as copper alloys.

Antifouling Properties of Copper Alloys

There is no biofouling on a copper alloy mesh after 4 months immersed in the waters of the North Atlantic (foreground), whereas hydroids have grown on high-density polyethylene tubing (background)

In the aquaculture industry, sound animal husbandry translates to keeping fish clean, well fed, healthy, and not overcrowded. One solution to keeping farmed fish healthy is to contain them in antifouling copper alloy nets and structures.

Researchers have attributed copper's resistance to biofouling, even in temperate waters, to two possible mechanisms: 1) a retarding sequence of colonization through release of antimicrobial copper ions, thereby preventing the attachment of microbial layers to marine surfaces; and 2) separating layers that contain corrosive products and the spores of juveniles or macro-encrusting organisms.

The most important requirement for optimum biofouling resistance is that the copper alloys should be freely exposed or electrically insulated from less noble alloys and from cathodic protection. Galvanic coupling to less noble alloys and cathodic protection prevent copper ion releases from surface films and therefore reduce biofouling resistance.

As temperatures increase and water velocities decrease in marine waters, biofouling rates dramatically rise. However, copper's resistance to biofouling is observed even in temperate waters. Studies in La Herradura Bay, Coquimbo, Chile, where biofouling conditions are extreme, demonstrated that a copper alloy (90% copper, 10% nickel) avoided macro-encrusting organisms.

Corrosion Behavior of Copper Alloys

Copper alloys used in sea water service have low general corrosion rates but also have a high resistance to many localized forms of corrosion. A technical discussion regarding various types of corrosion, application considerations (e.g., depth of installations, effect of polluted waters, sea conditions), and the corrosion characteristics of several copper alloys used in aquaculture netting is available (i.e., copper-nickel, copper-zinc, and copper-silicon).

Early Examples of Copper Sheathing

Prior to the late 1700s, hulls were made almost entirely of wood, often white oak. Sacrificial planking was the common mode of hull protection. This technique included wrapping a protective 1/2-inch thick layer of wood, often pine, on the hull to decrease the risk of damage. This layer was replaced regularly when infested with marine borers. Copper sheathing for bio-resistant ship hulls was developed in the late 18th century. In 1761, the hull of the British Royal Navy's HMS Alarm frigate was fully sheathed in copper to prevent attack by Teredo worms in tropical waters. The copper reduced biofouling of the hull, which enabled ships to move faster than those that did not have copper sheathed hulls.

Types of Copper Alloys

Copper–zinc brass alloys are currently being deployed in commercial-scale aquaculture operations in Asia, South America and the US (Hawaii). Extensive research, including demonstrations and trials, are currently being implemented on two other copper alloys: copper-nickel and copper-silicon. Each of these alloy types has an inherent ability to reduce biofouling, pen waste, disease, and the need for antibiotics while simultaneously maintaining water circulation and oxygen requirements. Other types of copper alloys are also being considered for research and development in aquaculture operations.

Section of a fish net on a salmon farm near Puerto Montt, Chile. The copper alloy woven mesh inside the frame has resisted biofouling whereas PVC (i.e., the frame around the mesh) is heavily fouled

The University of New Hampshire is in the midst of conducting experiments under the auspices of the International Copper Association (ICA) to evaluate the structural, hydrodynamic, and antifouling response of copper alloy nets. Factors to be determined from these experiments, such as drag, pen dynamic loads, material loss, and biological growth – well documented for nylon netting but not fully understood for copper-nickel alloy nets – will help to design fish pen enclosures made from these alloys. The East China Sea Fisheries Research Institute, in Shanghai, China, is also conducting experimental investigations on copper alloys for ICA.

Copper–Zinc Alloys

The Mitsubishi-Shindoh Co., Ltd., has developed a proprietary copper-zinc brass alloy, called UR30, specifically designed for aquaculture operations. The alloy, which is composed of 64% copper, 35.1% zinc, 0.6% tin, and 0.3% nickel, resists mechanical abrasion when formed into wires and fabricated into chain link, woven, or other types of flexible mesh. Corrosion rates depend on the depth of submersion and seawater conditions. The average reported corrosion rate reported for the alloy is < 5 μm/yr based on two- and five-year exposure trials in seawater.

The Ashimori Industry Company, Ltd., has installed approximately 300 flexible pens with woven chain link UR30 meshes in Japan to raise Seriola (i.e., yellowtail, amberjack, kingfish, hamachi). The company has installed another 32 brass pens to raise Atlantic salmon at the Van Diemen Aquaculture operations in Tasmania, Australia. In Chile, EcoSea Farming S.A. has installed a total of 62 woven chain link brass mesh pens to raise trout and Atlantic salmon. In Panama, China, Korea, Turkey, and the US, demonstrations and trials are underway using flexible pens with woven chain link UR30 and other mesh forms and a range of copper alloys.

To date, in over 10 years of aquaculture experience, chain link mesh fabricated by these brass alloys have not suffered from dezincification, stress corrosion cracking, or erosion corrosion.

Copper–Nickel Alloys

Copper–nickel alloys were developed specifically for seawater applications over five decades ago. Today, these alloys are being investigated for their potential use in aquaculture.

Copper–nickel alloys for marine applications are usually 90% copper, 10% nickel, and small amounts of manganese and iron to enhance corrosion resistance. The seawater corrosion resistance of copper–nickel alloys results in a thin, adherent, protective surface film which forms naturally and quickly on the metal upon exposure to clean seawater.

The rate of corrosion protective formation is temperature dependent. For example, at 27 °C (i.e., a common inlet temperature in the Middle East), rapid film formation and good corrosion protection can be expected within a few hours. At 16 °C, it could take 2–3 months for the protection to mature. But once a good surface film forms, corrosion rates decrease, normally to 0.02–0.002 mm/yr, as protective layers develop over a period of years. These alloys have good resistance to chloride pitting and crevice corrosion and are not susceptible to chloride stress corrosion.

Copper–Silicon Alloys

Copper–silicon has a long history of use as screws, nuts, bolts, washers, pins, lag bolts, and staples in wooden sailing vessels in marine environments. The alloys are often composed of copper, silicon, and manganese. The inclusion of silicon strengthens the metal.

As with the copper–nickel alloys, corrosion resistance of copper–silicon is due to protective films that form on the surface over a period of time. General corrosion rates of 0.025–0.050mm have been observed in quiet waters. This rate decreases towards the lower end of the range over long-term exposures (e.g., 400–600 days). There is generally no pitting with the silicon-bronzes. Also there is good resistance to erosion corrosion up to moderate flow rates. Because copper–silicon is weldable, rigid pens can be constructed with this material. Also, because welded copper–silicon mesh is lighter than copper-zinc chain link, aquaculture enclosures made with copper–silicon may be lighter in weight and therefore a potentially less expensive alternative.

References

- What-is-the-importance-of-a-fish-ladder: worldatlas.com, Retrieved 29 June, 2019
- "Ralco Acquires Patent for Intensive Raceway Shrimp Farming". Retrieved 2017-11-15
- Michael B. Timmons and James B. Ebeling (2013). Recirculating Aquaculture (3rd ed.). Ithaca Publishing Company Publishers. P. 3. ISBN 978-0971264656
- Boutwelluc, Juanita (December 15, 2007). "Aztecs' aquaponics revamped". Napa Valley Register. Archived from the original on December 20, 2013. Retrieved April 24, 2013
- Braithwaite, RA; mcevoy, LA (2005). "Marine biofouling on fish farms and its remediation". Advances in marine biology. 47: 215–52. Doi:10.1016/S0065-2881(04)47003-5. PMID 15596168
- "Aquaculture Netting by Industrial Netting". Industrialnetting.com. Archived from the original on 2010-05-29. Retrieved 2010-06-16
- Braithwaite, RA; mcevoy, LA (2005). Marine biofouling on fish farms and its remediation. Advances in Marine Biology. 47. Pp. 215–52. Doi:10.1016/S0065-2881(04)47003-5. ISBN 9780120261482. PMID 15596168
- Alessandra Bianchi (28 April 2009). "The next seafood frontier: The open ocean – Apr. 28, 2009". Money.cnn.com. Retrieved 16 June2010

Fish Farming 5

- Fish Hatchery
- Fish Stocking
- Aquaculture of Catfish
- Aquaculture of Cobia
- Aquaculture of Salmonids
- Aquaculture of Tilapia
- Broodstock
- Commercial Fish Feed
- Fish Disease and Parasites
- Fish Farming Issues

The practice of raising fish in tanks or enclosures primarily for food is known as fish farming. Some of the commonly farmed fish are catfish, cobia, tilapia and salmonids. This chapter discusses in detail the aquaculture of these fish as well as the various aspects of fish farming such as fish hatchery and fish stocking.

A fish farm is similar to a fish hatchery in that both can contain 500,000 and more fish. But, a fish hatchery is designed to raise the fish only to a young age before they are released into the wild, usually to bolster the numbers of that species. In contrast, a fish farm is designed to raise the fish until they are a size and age that makes them the best commercial value. The fish are ultimately retrieved and sold, typically as whole or processed food.

Fish farming is the most common form of aquaculture, and commonly involves trout, salmon, tilapia, cod, carp, and catfish. For a species such as cod, whose numbers in the Grand Banks fishery off the east coast of the Canadian maritime provinces plummeted to near zero in the 1970s due to overfishing, and as of 2008 have yet to recover, the cod available from fish farming represents almost the sole source of the fish in North American markets.

The example of cod is cited as one of the advantages of fish farming. Raising fish under more controlled conditions that are possible in the wild avoids the problem of overfishing. As well, because an operation takes up relatively little space, feeding and care of the fish can be done under more controlled conditions, which is an economic advantage to those who own and run the facility.

However, fish farming is a controversial practice. For example, on the Canadian west coast, the farming of salmon typically uses species normally found in the Atlantic Ocean. The escape of fish to the wild does occur, and has created concern that the presence of the species in an environment that is unnatural to them could upset the marine ecology. Other concerns of fish farming are the overcrowding of fish, which can make them more susceptible to disease such as sea lice, and the use of antibiotics, which can also be released into the natural environment.

Fish farming is an ancient practice, dating back to about 2500 BC in China, when carp were raised in ponds and in artificial lakes created by receding floodwaters. Some of the motivations for fish farming in ancient China are shared with fish farm owners and operators in 2008. These include maximizing the food available from the resource; reducing the energy needed to search for, gather, and transport the food; making food production more predictable and less likely to be influenced by weather, predators, or other factors; and ensuring that the quality of the resource remains acceptable over time.

Evidence of fish farming also dates back at least 1,000 years in Hawaii, when rocks were added to existing reefs to create an artificial pond. The spaces between the rocks were large enough to let seawater circulate in and out, allowing nutrients to circulate in and waste out, but were too small to allow the fish to escape. The open net design of present-day fish farms follows this example.

Fish farming became more prevalent in Europe in the fifteenth century. The first known fish hatchery constructed in North America was built in the Canadian province of Newfoundland in 1889.

In the 1960s, fish farming expanded worldwide as some commercial fish stocks became less plentiful and a growing global population increased the demand for fish. As with factory farms—the landlocked facilities in which huge numbers of poultry and livestock are raised—economic incentives were provided to encourage the establishment of freshwater and marine water fish farms. In addition, corporations involved in the sales of fresh and processed fish products began to expand into fish farming as a way of ensuring supply, expanding their market, and trimming costs.

As practiced in China thousands of years ago, fish farming was efficient and sustainable. The numbers of fish were suitable for the space available and the population was managed so that the numbers of fish ready for harvesting did not decline over time. When done in such a sustainable way, fish farming can be a good strategy to supplement or even replace the fish caught from the wild.

However, the confinement of large numbers of fish in a small area can create problems. In an enclosed pond, one problem can be the accumulation of waste products. Aside from making the water less hospitable for the fish, the waste material can act as a food source for microorganisms known as algae. Combined with suitable water temperature and sunlight, the presence of the food source can lead to the rapid increase in the number of algae termed as algal blooms. The number of algae in blooms that occur in the open ocean can be so large that the growth is visible from orbiting satellites. In a confined pond, an algal bloom can use up much of the oxygen in the water, leading to the death of the fish.

When a fish farm is done in tanks (closed circulation type), it is essential to keep the water well oxygenated and to remove wastes. Bubbling air into the water as is done with a home aquarium can be one means of oxygenation. Alternatively, water can be cascaded from tank to tank in the fish farm complex, with oxygen being supplied as the water tumbles between tanks.

Impacts and Issues

Fish farming has become a very contentious practice, for a number of environmental reasons and for the adverse health effects it has on the farmed fish and possibly other species, including humans.

In a fish farm, the concentration of fish far exceeds that found in schools of fish in the wild—50,000 or more fish in an area of several acres in volume—with the possible exception of the spawning runs of west coast salmon. These crowded conditions reduce the free-swimming volume of each fish to about that of the average household bathtub. In such crowded conditions, the fish bump and rub against each other in the boundaries of the pens, which can produce cuts and scrapes. This increases the likelihood of infection and the development of diseases.

Species of sea lice that parasitize Coho and Atlantic salmon are especially troublesome. The sea lice attach to the fish and feed on tissue, which creates lesions and causes fluid loss from the affected fish. The confined fish become ill and can die. In addition, the sea lice can spread to wild salmon in the seas around fish farms when farmed salmon escape from the confinement, and also when the lice are washed away from the fish farm into the surrounding water. A 2001 survey of wild juvenile salmon migrating past fish farms in British Columbia found many more sea lice on the juveniles that had passed the farms than on those who had not yet passed by the facilities.

The escape of fish from fish farms is not a trivial and isolated event. Rips and breaks in the pen material and buffeting of the pens by storm-driven waves can lead to the escape of fish. In some cases, pens are designed with a net lid to reduce this possibility. Sometimes only a few fish escape. But mass escapes have occurred. For example, in January 2002, over 8,000 fish escaped from a fish farm in Clayoquot Sound, British Columbia. Worldwide in 2004, an estimated 2 million farmed fish escaped to the wild.

Once in the wild, the farmed fish have the potential to transfer disease to the wild population. An article in a December 2007 issue of Science documented declines in the population of wild Pacific salmon related to their decimation from sea lice transferred from farm populations of Atlantic salmon. The situation is so dire that the natural population could be reduced by 99% by 2015, which would be an economic disaster for the traditional salmon fishery and those employed by the fishery.

Antibiotics can be supplied to the food in an effort to control infections. As with land-bound factory farms, this practice encourages the development of antibiotic resistance among the surviving bacteria. These hardier varieties of the bacteria may pose a health hazard not only to the farmed fish, but to wild fish populations and to humans.

A fish farm releases a great amount of untreated sewage to the surrounding water. A study done in Clayoquot Sound calculated that the 700,000 fish housed in the facility that is the size of three football fields generate the daily equivalent amount of sewage produced by 150,000 people.

Elsewhere, the situation is not better. In China, for example, which produces approximately 70% of the world's farmed fish, fish farms can be concentrated together around large ponds. Analysis of the pond waters has revealed the presence of pesticides and other agricultural run-off, antibiotics, and cancer-causing compounds. The result is both an environment and food safety problem. In late 2007, the United States and China signed an agreement to permit more monitoring of the farms and the safety of the exported products.

Fish Hatchery

A fish hatchery is a place for artificial breeding, hatching, and rearing through the early life stages of animals—finfish and shellfish in particular. Hatcheries produce larval and juvenile fish, shellfish, and crustaceans, primarily to support the aquaculture industry where they are transferred to on-growing systems, such as fish farms, to reach harvest size. Some species that are commonly raised in hatcheries include Pacific oysters, shrimp, Indian prawns, salmon, tilapia and scallops. The value of global aquaculture production is estimated to be US$98.4 billion in 2008 with China significantly dominating the market; however, the value of aquaculture hatchery and nursery production has yet to be estimated. Additional hatchery production for small-scale domestic uses, which is particularly prevalent in South-East Asia or for conservation programmes, has also yet to be quantified.

There is much interest in supplementing exploited stocks of fish by releasing juveniles that may be wild caught and reared in nurseries before transplanting, or produced solely within a hatchery. Culture of finfish larvae has been utilised extensively in the United States in stock enhancement efforts to replenish natural populations. The U.S. Fish and Wildlife Service have established a National Fish Hatchery System to support the conservation of native fish species.

Purpose

Assynt Salmon hatchery, near Inchnadamph in the Scottish Highlands

Hatcheries produce larval and juvenile fish and shellfish for transferral to aquaculture facilities where they are 'on-grown' to reach harvest size. Hatchery production confers three main benefits to the industry:

1. Out of Season Production:

Consistent supply of fish from aquaculture facilities is an important market requirement. Broodstock conditioning can extend the natural spawning season and thus the supply of juveniles to farms. Supply can be further guaranteed by sourcing from hatcheries in the opposite hemisphere i.e. with opposite seasons.

2. Genetic Improvement:

Genetic modification is conducted in some hatcheries to improve the quality and yield of farmed species. Artificial fertilisation facilitates selective breeding programs which aim to improve production characteristics such as growth rate, disease resistance, survival, colour, increased fecundity and lower age of maturation. Genetic improvement can be mediated by selective breeding, via hybridization, or other genetic manipulation techniques.

3. Reduce dependence on wild-caught juveniles:

In 2008 aquaculture accounted for 46% of total food fish supply, around 115 million tonnes. Although wild caught juveniles are still utilised in the industry, concerns over sustainability of extracting juveniles, and the variable timing and magnitude of natural spawning events, make hatchery production an attractive alternative to support the growing demands of aquaculture.

Production Steps

Manually stripping eggs

Juvenile salmon towards the end of their stay in a hatchery

Broodstock

Broodstock conditioning is the process of bringing adults into spawning condition by promoting the development of gonads. Broodstock conditioning can also extend spawning beyond natural spawning periods, or for production of species reared outside their natural geographic range with different environmental conditions. Some hatcheries collect wild adults and then bring them in for conditioning whilst others maintain a permanent breeding stock. Conditioning is achieved by holding broodstock in flow-through tanks at optimal conditions for light, temperature, salinity, flow rate and food availability (optimal levels are species specific). Another important aspect of broodstock conditioning is ensuring the production of high quality eggs to improve growth and survival of larvae by optimising the health and welfare of broodstock individuals. Egg quality is often determined by the nutritional condition of the mother. High levels of lipid reserves in particular are required to improve larval survival rates.

Spawning

Natural spawning can occur in hatcheries during the regular spawning season however where more control over spawning time is required spawning of mature animals can be induced by a variety of methods. Some of the more common methods are:

- Manual stripping: For shellfish, gonads are generally removed and gametes are extracted or washed free. Fish can be manually stripped of eggs and sperm by stroking the anaesthetised fish under the pectoral fins towards the anus causing gametes to freely flow out.

- Environmental manipulation: Thermal shock, where cool water is alternated with warmer water in flow-through tanks can induce spawning. Alternatively, if environmental cues that stimulate natural spawning are known, these can be mimicked in the tank e.g. changing salinity to simulate migratory behaviour. Many individuals can be induced to spawn this way, however this increases the likelihood of uncontrolled fertilisation occurring.

- Chemical injection: A number of chemicals can be used to induce spawning with various hormones being the most commonly used.

Fertilisation

Prior to fertilisation, eggs can be gently washed to remove wastes and bacteria that may contaminate cultures. Promoting cross-fertilisation between a large number of individuals is necessary to retain genetic diversity in hatchery produced stock. Batches of eggs are kept separate, fertilised with sperm obtained from several males and allowed to stand for an hour or two before samples are analyzed under a microscope to ensure high rates of fertilisation and to estimate numbers to be transferred to larval rearing tanks.

Larvae

Rearing larvae through the early life stages is conducted in nurseries which are generally closely associated with hatcheries for fish culture whilst it is common for shellfish nurseries to exist separately. Nursery culture of larvae to rear juveniles of a size suitable for transferral to on-growing facilities can be performed in a variety of different systems which may be entirely land-based, or larvae may be later transferred to sea-based rearing systems which reduce the need to supply feed. Juvenile survival is dependent on very high quality water conditions. Feeding is an important component of the rearing process. Although many species are able to grow on maternal reserves alone (lecithotrophy), most commercially produced species require feeding to optimise survival, growth, yield and juvenile quality. Nutritional requirements are species specific and also vary with larval stage. Carnivorous fish are commonly fed with live prey; rotifers are usually offered to early larvae due to their small size, progressing to larger *Artemia* nauplii or zooplankton. The production of live feed on-site or buying-in is one of the biggest costs for hatchery facilities as it is a labour-intensive process. The development of artificial feeds is targeted to reduce the costs involved in live feed production and increase the consistency of nutrition, however decreased growth and survival has been found with these alternatives.

Settlement of Shellfish

The hatchery production of shellfish also involves a crucial settling phase where free-swimming larvae settle out of the water onto a substrate and undergo metamorphosis if suitable conditions are found. Once metamorphosis has taken place the juveniles are generally known as spat, it is this phase which is then transported to on-growing facilities. Settlement behaviour is governed by a range of cues including substrate type, water flow, temperature, and the presence of chemical cues indicating the presence of adults, or a food source etc. Hatchery facilities therefore need to understand these cues to induce settlement and also be able to substitute artificial substrates to allow for easy handling and transportation with minimal mortality.

Hatchery Design

Multi-Species Fish and Invertebrate Breeding and Hatchery

Hatchery designs are highly flexible and are tailored to the requirements of site, species produced, geographic location, funding and personal preferences. Many hatchery facilities are small and coupled to larger on-growing operations, whilst others may produce juveniles solely for sale. Very small-scale hatcheries are often utilized in subsistence farming to supply families or communities particularly in south-east Asia. A small-scale hatchery unit consists of larval rearing tanks, filters, live food production tanks and a flow through water supply. A generalized commercial scale hatchery would contain a broodstock holding and spawning area, feed culture facility, larval culture area, juvenile culture area, pump facilities, laboratory, quarantine area, and offices and bathrooms.

Expense

Labour is generally the largest cost in hatchery production making up more that 50% of total costs. Hatcheries are a business and thus economic viability and scale of production are vital considerations. The cost of production for stock-enhancement programmes is further complicated by the difficulty of assessing the benefits to wild populations from restocking activities.

Issues

Genetic

Hatchery facilities present three main problems in the field of genetics. The first is that maintenance of a small number of broodstock can cause inbreeding and potentially lead to inbreeding depression thus affecting the success of the facility. Secondly, hatchery reared juveniles, even from a fairly large broodstock, can have greatly reduced genetic diversity compared to wild populations (the situation is comparable to the founder effect). Such fish that escape from farms or are released for restocking purposes may adversely affect wild population genetics and viability. This is of particular concern where escaped fish have been actively bred or are otherwise genetically modified. The third key issue is that genetic modification of food items is highly undesirable for many people.

Fish Farms

Other arguments that surround fish farms such as the supplementation of feed from wild caught species, the prevalence of disease, fish welfare issues and potential effects on the environment are also issues for hatchery facilities.

Fish Stocking

Fish stocking is the practice of raising fish in a hatchery and releasing them into a river, lake, or ocean to supplement existing populations or to create a population where none exists. Stocking may be done for the benefit of commercial, recreational, or tribal fishing, but may also be done to restore or increase a population of threatened or endangered fish in a body of water closed to fishing.

Its may be conducted by governmental agencies in public waters, or by private groups in private waters. When in public waters, fish stocking creates a common-pool resource which is rivalrous in nature but non-excludable. Thus, on public grounds, all can enjoy the benefits of fishing so long as fish continue to be stocked.

Fish stocking is a practice that dates back hundreds of years. According to biologist Edwin Pister, widespread trout stocking dates back to the 1800s. For the first hundred years of stocking, the location and number of fish introduced was not well recorded; the singular goal of stocking was to enhance sport fishing regardless of ecological ramifications such as erosion of biodiversity. As biologist Edwin Pister states, "When trout planting was first implemented, the nation was gripped with a highly utilitarian resource management ethic that placed short-term human interests above virtually any other consideration". Recently, the U.S. Fish and Wildlife Service along with state fishery branches have done a better job of recording exactly what species of fish are stocked at any given location. This began in the 1960s when research suggested the negative impacts of fish stocking on the ecological complexity of other life forms. The Wilderness Act of 1964 also opened the public's eyes to the impact stocking has on other organisms. Thus, fish stocking is now the subject of much debate as there are various costs and benefits associated with the practice.

In the United States, stocking non-native fish for sport and food was just beginning in 1871 when the US Fish Commission was established. The head of the new agency, Spencer Fullerton Baird, was tasked to research "the decrease of the food fishes of the seacoasts and the lakes of the United States and to suggest remedial measures." Baird made his headquarters at Woods Hole on Cape Cod, Massachusetts. There, his team of scientists and researchers conducted studies on striped bass, blue fish, and many other commercial and sport fish. They compiled their research into a 255-page report on United States fish resources. Congress granted the team $15,000 to develop food fish stocks, and nonnative fish such as rainbow trout, salmon, striped bass, and carp were subsequently introduced successfully into United States lakes and rivers. In the early years, fish were stocked by sports clubs and private citizens. Today, state fish and wildlife agencies along with hatcheries are responsible for distributing fish. And until recently, their goal was to plant as many fish as possible into as many bodies of water as possible. Now, with knowledge of the detrimental effects fish stocking has on invertebrate and amphibian populations, it is conducted much more selectively.

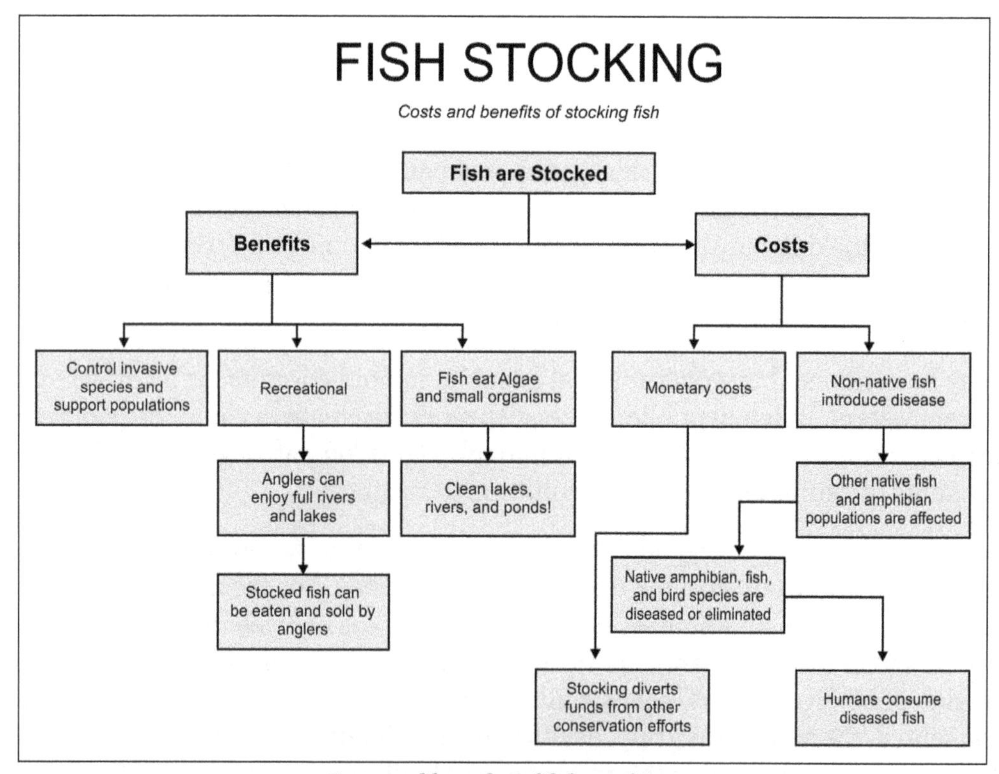

Costs and benefits of fish stocking

Today, much more thought is put into introducing non-native species as they can severely damage the populations of fragile natives; practices lean more towards sustainability.

Stocking is used to restore native species to waters where they have been overfished or can no longer breed. "Give and take" stocking practices are those where fish are stocked only to be fished and then restocked. In response, most states have adopted regulations that prohibit fish stocking in areas that may damage aquatic life or ecosystem diversity, and encourage stocking in bodies of water where no harm will result from doing so. Trout Unlimited has a policy, for example, that states "where a body of scientific evidence shows that stocking in historically non-salmonid waters adversely affects native biodiversity, such stocking should cease." While many organizations remain focused solely on providing quality fishing opportunities, policies and attitudes are shifting toward resource integrity and protection.

In Pennsylvania alone, the PA Fish & Boat Commission is scheduled to stock 4,398,227 trout (brook, brown, and rainbows) into its streams and lakes in 2019. Also in 2019, Lake Ontario, one of the five great lakes, is projected to receive 2,767,660 stocked salmon and trout. According to data by state agencies, in 2004 roughly 1.7 billion fish were stocked across the United States. With 104 different species of fish stocked, a total of 43.65 million pounds of fish were released, primarily in the Western states. In the US, common species that are currently stocked for sport include trout, bass, salmon, muskellunge, walleye, and several species of panfish.

Before being released into rivers, lakes, ponds, and occasionally oceans across the country, fish are raised in a fish hatchery. Just like humans have a demand for groceries, anglers have a demand for fish. When the supply is running low in a given body of water, fish are transported from hatcheries in a large water tank or airplane to their respective locations. The costs associated with stocking are typically covered by angler associations, commercial fishermen, state fish and game agencies, and at times government subsidies; today, most stocking is conducted by state fish and game departments.

Benefits of Stocking

Stocking fish provides a variety of benefits to society. Fishing is a popular pastime, with 101.6 million Americans over 16 years old participating in wildlife-related activities such as fishing and hunting. Additionally, a report by the U.S. Department of the Interior recorded an 8% increase in angling participation between 2011 and 2016. In 2017, just over 49 million Americans participated in some form of fishing, of which 11.9 million were youth. Most anglers even fish between four and eleven times per year, with some reporting over 100 trips. Adding fish to public lakes and streams makes fishing more fun, especially for young anglers. Fishing also provides Americans the opportunity to enjoy nature in solitude and escape from their busy lives. It is a form of exercise and a great way to bond with friends and family; fishing has also been linked to increased patience among children. According to a survey by the Outdoor Foundation, 60.3% of fisherman report fishing as a means of getting exercise, 59.1% report fishing to bond with family, and 51.2% report fishing to be close to nature and observe the scenery. Some of the fish caught are also large, providing a source of nutrition to successful fishermen. Roughly 40% of fisherman eat the fish they catch, according to a survey by the Outdoor Foundation. Stocking can also restore threatened, endangered, or native fish species. The Union for Conservation of Nature's red list of endangered species has 1,414 species of fish that are at risk of extinction. Stocking them into lakes, rivers, and streams can support existing populations that are threatened and reduce the number of endangered or extirpated species. Many of the fish commonly used for stocking also have low reproductive rates and tend to be overfished if not stocked annually. Moreover, as stocked fish tend to contain larger trophy fish, many anglers are more willing to pay for a fishing license, meaning state fishing departments have more revenue to spend on natural resource management and conservation efforts. In 2018, there were roughly 30 million paid license holders in the US, grossing $720 million that year.

Fishing for stocked trout in Wyoming

Many species of fish including grass carp and the suckermouth catfish help clean bodies of water by eating algae and other green organisms. Algae can take over stagnant ponds, attracting insects and making lakes, rivers, and ponds unpleasant to look at. To treat them, many individuals will choose to stock certain species of fish. This creates a positive externality for those who enjoy a variety of water activities. Studies have also examined the economic viability of fish stocking. Hansson, Arrheniusm, and Nellbring of Stockholm University find that simple economic analysis suggests Volga pikeperch stocking can be profitable; based on the capital invested in the stock, the economic yield results in an annual interest rate of 43% (from the viewpoint of anglers). These authors also find that increased populations of stocked fish decreases manpower and equipment costs associated with each catch. They also find that specifically for pikeperch, stocking can restore food web interactions to a more "natural" level where herring populations are reduced and zooplankton flourish, thus benefiting the ecosystem as a whole.

Drawbacks and Risks

Although there are many benefits of stocking, some groups, including the Pacific Rivers Council, believe that it diverts money away from more effective conservation and protection efforts. In addition, declining populations of native Golden Trout, a now threatened species, has been linked to stocking of nonnative species. As a matter of fact, 35 species of fish and amphibians have been negatively affected by stocking practices in California. Scientists have established a direct link between nonnative fish stocking and these declines: golden trout, Lahontan cutthroat trout, mountain yellow-legged frog, Yosemite toad, and Cascades frog are all threatened by the practice. Mating between native and introduced fish species can lower the fitness of natural populations, thereby introducing diseases that affect fish and other wildlife. In fact, a study conducted in Virginia streams found that an infectious virus was found only in brook trout populations that had a history of stocking. These diseases may affect humans who consume them as well. Maintaining a balanced ecosystem with biological diversity is also extremely important. Many scientists have claimed that because fish that are stocked tend to be apex predators, native species may become prey and have to compete with the oftentimes larger introduced fish for food and cover. Additionally, the use of aircraft to stock fish in the second half of the twentieth century meant pilots often stocked the wrong lakes or rivers. In many cases, this had disastrous consequences. When fish, specifically trout, are stocked into ecologically sensitive bodies of water, invertebrate populations and amphibians are threatened, altering the natural selection pressures within the ecosystem. Stocked rainbow trout have been outcompeting native brook trout in many southeastern United States bodies of water, for example. Even bird populations such as cormorants are affected. According to findings by the Ecological Society of America, when gulls in the Great Lakes area were examined after fish stocking, they consumed more garbage, presumably due to the decrease in native fish by the predatory stocked fish. Another study found that stocked fish in the Pacific Northwest spread a disease that caused a 15% increase in amphibian embryo mortality. In 2005, the Center for Biological Diversity studied bodies of water across California and found that nonnative trout had been stocked in at least 47 areas where rare species were present, damaging 39 imperiled fish and amphibians. Members of the Wilderness Research Institute claim that fish stocking compromises the "naturalness" of aquatic ecosystems and that protecting water quality is more important than recreational opportunities. Although stocking programs were designed to boost conservation, in some cases they have harmed them. The increased boat traffic associated with better fishing opportunities creates negative externalities. These include water contamination, congestion, noise pollution, and disruption of wilderness experiences.

Fish are stocked all over the world. This river in Aspiring National Park, NZ is stocked with trout

However, there are benefits to fish stocking. Anglers across the country spend millions of dollars through license fees (which benefit state governments) and fishing equipment such as rods, reels, and lures. Members of fishing societies such as the International Game Fish Association and the Bass Anglers Sportsman Society also enjoy fishing in more robust bodies of water. It is important to note that despite the findings mentioned above, Virginia researchers have found that even with stockings of rainbow trout, 80% of fish in its stocked streams are still natives. The Department of Natural Resources in Minnesota also found that stocked muskie can coexist because biologically-based guidelines are used when choosing what and where to stock. Until recently, the ecology of lakes and rivers was not well understood. In order to maximize the benefits of stocking and control the costs, fish must be stocked strategically in places where they can be enjoyed but do not pose a threat to native species.

How to Stock a Private Pond

When stocking a private pond, it is important to introduce three prey fish for every predator fish. This ensures that predatory fish have enough nutrition to survive and that prey fish can still reproduce. It is also recommended to stock fathead minnows so that both predator and prey fish have enough to feed on. Introducing fish of similar sizes is another important step to ensure that the population grows together. At the same time, make sure that the existing fish in the pond are not significantly larger than the ones being added. For a standard one-quarter acre pond, it is recommended to stock 120 sunfish, 60 yellow perch, 15 largemouth bass, and 8 pounds of fathead minnows. One way to determine what kinds of fish are already living in a given body of water is to monitor local streams, rivers, and lakes and record what species of fish are being caught.

The best time to stock is in the spring or fall due to mild temperatures and higher levels of oxygen in the water. To acclimate the fish, place the transportation bag in a shaded part of the water and leave it floating for 15-20 minutes. Before releasing the fish, make sure that larger fish and smaller fish are released at different ends of the pond, giving the prey an opportunity to find shelter. Studies show that releasing small numbers of fish at regular intervals is more effective than releasing all at once, so if possible, plan to release them over a few weeks.

It is also important to stock the correct species of fish. For warm water ponds, it is recommended to stock largemouth bass, bluegill, crappie, channel catfish, and bullheads. For larger and deeper lakes, stocking cool water game fish such as walleye and trout species is recommended. Lastly, it is important to make sure that no pond or body of water is overstocked. Each has a carrying capacity, meaning that any given body of water can only sustain a certain amount of fish. If this carrying capacity is exceeded, fish will have to compete for food and cover, resulting in damage to all organisms in the water.

Aquaculture of Catfish

Catfish are easy to farm in warm climates, leading to inexpensive and safe food at local grocers. Catfish raised in inland tanks or channels are considered safe for the environment, since their waste and disease should be contained and not spread to the wild.

In Asia, many catfish species are important as food. Several walking catfish (Claridae) and shark catfish (Pangasiidae) species are heavily cultured in Africa and Asia. Exports of one particular shark catfish species from Vietnam, *Pangasius bocourti*, has met with pressures from the U.S. catfish industry. In 2003, The United States Congress passed a law preventing the imported fish from being labeled as catfish. As a result, the Vietnamese exporters of this fish now label their products sold in the U.S. as "basa fish".

Ictalurids are cultivated in North America (especially in the Deep South, with Mississippi being the largest domestic catfish producer). Channel catfish (*Ictalurus punctatus*) supports a $450 million/yr aquaculture industry. The US farm-raised catfish industry began in the early 1960s in Kansas, Oklahoma and Arkansas. Channel catfish quickly became the major catfish grown, as it was hardy and easily spawned in earthen ponds. By the late 60s, the industry moved into the Mississippi Delta as farmers struggled with sagging profits in cotton, rice and soybeans, especially on those farm areas where soils had a very high clay content.

The Mississippi Delta became the industry home for the catfish industry, as they had the soils, climate and shallow aquifers to provide water for the earthen ponds that grow 360-380 million pounds (160,000 to 170,000 tons) of catfish annually. Catfish are fed a grain-based diet that includes soybean meal. Fish are fed daily through the summer at rates of 1-6% of body weight with the pelleted floating feed. Catfish need about two pounds of feed to produce one pound of live weight. Mississippi is home to 100,000 acres (400 km^2) of catfish ponds, the largest of any state. Other states important in growing catfish include Alabama, Arkansas and Louisiana.

There is a large and growing ornamental fish trade, with hundreds of species of catfish, such as *Corydoras* and armored suckermouth catfish (often called plecos), being a popular component of many aquaria. Other catfish commonly found in the aquarium trade are banjo catfish, talking catfish, and long-whiskered catfish.

Aquaculture of Cobia

Cobia, a warm water fish, is one of the more suitable candidates for offshore aquaculture. Cobia are large pelagic fish, up to 2 metres (78 inches) long and 68 kilograms (150 pounds) in weight. They are solitary fish except when spawning, found in warm-temperate to tropical waters.

Their rapid growth rate in aquaculture, as well as the high quality of their flesh, makes cobia potentially one of the more important potential marine fish for aquaculture production. Currently, cobia are cultured in nurseries and grow-out offshore cages in many parts of Asia and off the coast of the United States, Mexico and Panama. In Taiwan cobia weighing 100–600 grams are cultured for 1–1.5 years to reach the 6–8 kilograms needed for export to Japan. Currently, around 80% of marine cages in Taiwan are devoted to cobia culture. In 2004, the FAO reported that 80.6% of the world's cobia production was by China and Taiwan. After China and Taiwan, Vietnam is the third largest producer of farmed cobia in the world where production was estimated at 1500 tonnes in 2008. The possibility is also being examined of growing hatchery reared cobia in offshore cages around Puerto Rico and the Bahamas.

Greater depths, stronger currents, and distance from shore all act to reduce the environmental impacts often associated with fin fish aquaculture. Offshore cage systems could become some of the most environmentally sustainable methods for commercial marine fish aquaculture. However, some problems still exist in cobia culture that needs to be addressed and solved for increasing production. These include high mortality rates due to stress during transport from nursery tanks or inshore cages out to grow-out cages. Also, diseases in the nursery stage and the grow-out culture can result in low survival rates and a poor harvest.

Production

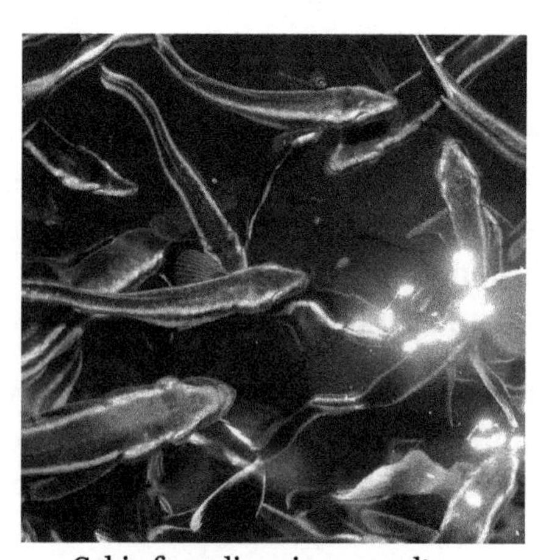

Cobia fingerlings in aquaculture

Wild cobia broodstock are captured by professional fishermen. The fish are transferred into onboard-tanks on a transport vessel for transport to hatchery facilities. They are anesthetized with clove oil if necessary to reduce stress during transportation. They are also treated for ectoparasites on their gills and skin that could proliferate later after transfer to maturation tanks.

Broodstock are reared in controlled ponds or tanks. These tanks are often stocked with cleaner fish, *Gobiosoma oceanops*, as a biological control for any remaining ectoparasites.

The broodstock diet includes sardines, squid and formulated feeds, as well as vitamin and mineral supplements. The water temperature is used to control spawning.

The eggs are collected with a surface skimmer using mesh screen bags. The eggs are transferred to incubation tanks where they are disinfected for an hour with 100 ppm formalin.

Phytoplankton concentrations are maintained, and enriched *Artemia* nauplii and rotifers are fed to the cobia larvae for 3–7 days after they hatch. The larvae require rotifers for at least four days after hatching. The presence of enriched live prey in conjunction with live algae in rearing tanks has been shown to improve the way larvae grow and survive in recirculating systems.

Optimal rearing densities are required when rearing larvae. Even though water quality and food can be controlled, it has been shown that high rearing densities may still affect growth and survival of the larvae through responses related to crowding. In addition, juveniles exposed to varying salinities exhibited sustained growth and improved health at higher salinities, 15 and 30 ppt.

Cobia larvae metamorphose to gill respiration 11–15 days post hatching. At 15–25 days post hatching, cobia are weaned onto commercial formulated feeds. Rearing cobia larvae at salinities as low as 15 ppt is possible. Fully weaned fingerlings weighing up to one gram are transferred to juvenile culture tanks. Later cobia juveniles can be raised in ponds or shallow, near-shore submerged cages.

Juveniles thrive on a wide range of protein and lipid, but there are optimal levels where they get the most benefit. After an 8-week growth trial, juvenile cobia displayed a peak in their weight gain with a dietary protein concentration of 44.5%. Weight gain is also likely to increase as the lipid content in the diet increases. However, levels exceeding 15–18% produces little practical benefit because of higher fat accretion in the cobia. In addition, up to 40% of fish meal protein can be substituted with soybean meal protein before a reduction occurs in growth rates and protein utilization. Cobia has low feed conversion rates, yielding 1 kilogram of fish biomass for 1.8 kilograms of pellets which contain 50% fishmeal.

The cobia are then transferred to open ocean cages for final the grow-out when they reach 6–10 kilograms. The growth rate and survival rate of cobia during grow-out stages in open water cages throughout the Caribbean and Americas vary from as little as 10% up to 90%. Low survival rates are mainly due to disease, but also to shark attacks which tear holes in the nets of cages in the Bahamas and Puerto Rico and allow caged cobia to escape. However, better growth rates were experienced in offshore cage farms in Taiwan. In addition, cobia are considered to be gonochoristic, with differential growth rates occurring between sexes. Females grow faster and have been shown to be significantly longer and heavier within year classes.

Diseases

- Nephrocalcinosis (kidney stones) cause significant mortality during both the hatchery and grow-out stages. These stones vary in diameter from 2–6 mm in the kidney and can block the urethra. This condition is not fully understood, but is thought to be a symptom of prolonged exposure to free carbon dioxide in excess of 10 mg/L. The ratio of calcium to magnesium in the diet could also be out of balance.

- A Sphaerospora-like myxosporean infection caused 90% mortality during one month in a marine cage cultured in Taiwan.

Benefits and Constraints

Offshore aquaculture, regardless of the species, is beneficial because it can avoid conflict with recreational activities and local fisherman, as well as potentially improving the coastal aesthetics. Further, repositioning aquaculture facilities in less polluted open water environments can produce better products, and the high flushing rates experienced in the open ocean reduces the effect of effluents on benthic communities.

However, such operations require more developed infrastructure than near-shore aquaculture systems, which makes them expensive. Offshore sites have access difficulties and much higher labour costs.

Aquaculture of Salmonids

The aquaculture of salmonids is the farming and harvesting of salmonids under controlled conditions for both commercial and recreational purposes. Salmonids (particularly salmon and rainbow trout), along with carp, and tilapia are the three most important fish species in aquaculture. The most commonly commercially farmed salmonid is the Atlantic salmon. In the U.S. Chinook salmon and rainbow trout are the most commonly farmed salmonids for recreational and subsistence fishing through the National Fish Hatchery System. In Europe, brown trout are the most commonly reared fish for recreational restocking. Commonly farmed nonsalmonid fish groups include tilapia, catfish, sea bass, and bream.

In 2007, the aquaculture of salmonids was worth US$10.7 billion globally. Salmonid aquaculture production grew over ten-fold during the 25 years from 1982 to 2007. In 2012, the leading producers of salmonids were Norway, Chile, Scotland and Canada.

Much controversy exists about the ecological and health impacts of intensive salmonids aquaculture. Of particular concern are the impacts on wild salmon and other marine life. Some of this controversy is part of a major commercial competitive fight for market share and price between Alaska commercial salmonid fishermen and the rapidly evolving salmonid aquaculture industry.

Methods

The aquaculture or farming of salmonids can be contrasted with capturing wild salmonids using commercial fishing techniques. However, the concept of "wild" salmon as used by the Alaska Seafood Marketing Institute includes stock enhancement fish produced in hatcheries that have historically been considered ocean ranching. The percentage of the Alaska salmon harvest resulting from ocean ranching depends upon the species of salmon and location. Methods of salmonid aquaculture originated in late 18th-century fertilization trials in Europe. In the late 19th century, salmon hatcheries were used in Europe and North America. From the late 1950s, enhancement programs based on hatcheries were established in the United States, Canada, Japan, and the USSR. The contemporary technique using floating sea cages originated in Norway in the late 1960s.

Assynt salmon hatchery, near Inchnadamph in the Scottish Highlands

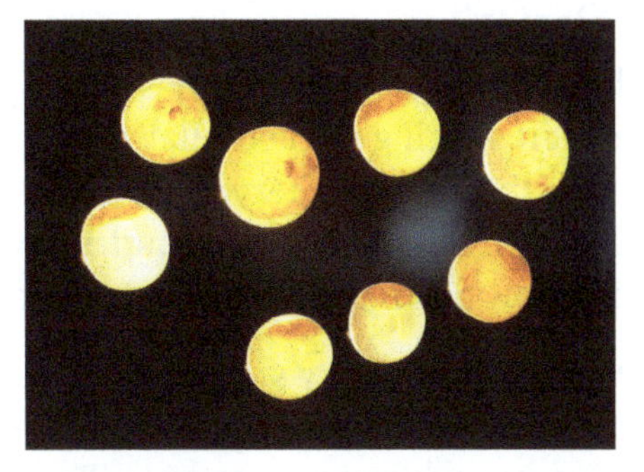

Very young fertilised salmon eggs, notice the developing eyes and vertebral column

Salmonids are usually farmed in two stages and in some places maybe more. First, the salmon are hatched from eggs and raised on land in freshwater tanks. Increasing the accumulated thermal units of water during incubation reduces time to hatching. When they are 12 to 18 months old, the smolt (juvenile salmon) are transferred to floating sea cages or net pens anchored in sheltered bays or fjords along a coast. This farming in a marine environment is known as mariculture. There they are fed pelleted feed for another 12 to 24 months, when they are harvested.

Salmon egg hatching: In about 24 hr, it will be a fry without the yolk sac

Norway produces 33% of the world's farmed salmonids, and Chile produces 31%. The coastlines of these countries have suitable water temperatures and many areas well protected from storms. Chile is close to large forage fisheries which supply fish meal for salmon aquaculture. Scotland and Canada are also significant producers; and it was reported in 2012 that the Norwegian government at that time controlled a significant fraction of the Canadian industry.

Modern salmonid farming systems are intensive. Their ownership is often under the control of huge agribusiness corporations, operating mechanized assembly lines on an industrial scale. In 2003, nearly half of the world's farmed salmon was produced by just five companies.

Hatcheries

Modern commercial hatcheries for supplying salmon smolts to aquaculture net pens have been shifting to recirculating aquaculture systems (RAS)s where the water is recycled within the hatchery. This allows location of the hatchery to be independent of a significant fresh water supply and allows economical temperature control to both speed up and slow down the growth rate to match the needs of the net pens.

Conventional hatchery systems operate flow-through, where spring water or other water sources flow into the hatchery. The eggs are then hatched in trays and the salmon smolts are produced in raceways. The waste products from the growing salmon fry and the feed are usually discharged into the local river. Conventional flow-through hatcheries, for example the majority of Alaska's enhancement hatcheries, use more than 100 tonnes (16,000 st) of water to produce a kg of smolts.

An alternative method to hatching in freshwater tanks is to use spawning channels. These are artificial streams, usually parallel to an existing stream with concrete or riprap sides and gravel bottoms. Water from the adjacent stream is piped into the top of the channel, sometimes via a header pond to settle out sediment. Spawning success is often much better in channels than in adjacent streams due to the control of floods which in some years can wash out the natural redds. Because of the lack of floods, spawning channels must sometimes be cleaned out to remove accumulated sediment. The same floods which destroy natural redds also clean them out. Spawning channels preserve the natural selection of natural streams as no temptation exists, as in hatcheries, to use prophylactic chemicals to control diseases. However, exposing fish to wild parasites and pathogens using uncontrolled water supplies, combined with the high cost of spawning channels, makes this technology unsuitable for salmon aquaculture businesses. This type of technology is only useful for stock enhancement programs.

Sea Cages

Sea cages, also called sea pens or net pens, are usually made of mesh framed with steel or plastic. They can be square or circular, 10 to 32 m (33 to 105 ft) across and 10 m

(33 ft) deep, with volumes between 1,000 and 10,000 m^3 (35,000 and 353,000 cu ft). A large sea cage can contain up to 90,000 fish.

They are usually placed side by side to form a system called a seafarm or seasite, with a floating wharf and walkways along the net boundaries. Additional nets can also surround the seafarm to keep out predatory marine mammals. Stocking densities range from 8 to 18 kg (18 to 40 lb)/m^3 for Atlantic salmon and 5 to 10 kilograms (11 to 22 lb)/m^3 for Chinook salmon.

In contrast to closed or recirculating systems, the open net cages of salmonid farming lower production costs, but provide no effective barrier to the discharge of wastes, parasites, and disease into the surrounding coastal waters. Farmed salmon in open net cages can escape into wild habitats, for example, during storms.

An emerging wave in aquaculture is applying the same farming methods used for salmonids to other carnivorous finfish species, such as cod, bluefin tuna, halibut, and snapper. However, this is likely to have the same environmental drawbacks as salmon farming.

A second emerging wave in aquaculture is the development of copper alloys as netting materials. Copper alloys have become important netting materials because they are antimicrobial (i.e., they destroy bacteria, viruses, fungi, algae, and other microbes), so thery prevent biofouling (i.e., the undesirable accumulation, adhesion, and growth of microorganisms, plants, algae, tubeworms, barnacles, mollusks, and other organisms). By inhibiting microbial growth, copper alloy aquaculture cages avoid costly net changes that are necessary with other materials. The resistance of organism growth on copper alloy nets also provides a cleaner and healthier environment for farmed fish to grow and thrive.

Feeding

Salmonids are carnivorous and are currently being fed compound fish feeds containing fish meal and other feed ingredients, ranging from wheat byproducts to soybean meal and feather meal. Being aquatic carnivores, salmonids do not tolerate or properly metabolize many plant-based carbohydrates and use fats instead of carbohydrates as a primary energy source.

With the amount of worldwide fish meal production being almost a constant amount for the last 30+ years and at maximum sustainable yield, much of the fish meal market has shifted from chicken and pig feed to fish and shrimp feeds as aquaculture has grown in this time.

Work continues on developing salmonid diet made from concentrated plant protein. As of 2014, an enzymatic process can be used to lower the carbohydrate content of barley, making it a high-protein fish feed suitable for salmon. Many other substitutions for fish meal are known, and diets containing zero fish meal are possible. For example, a planned closed-containment salmon fish farm in Scotland uses ragworms, algae, and amino acids as feed. Some of the eicosapentaenoic acid and docosahexaenoic acid (in Omega-3 fatty acids) may be replaced by land-based (non-marine) algae oil, reducing the harvest of wild fish as fish meal.

However, commercial economic animal diets are determined by least-cost linear programming models that are effectively competing with similar models for chicken and pig feeds for the same feed ingredients, and these models show that fish meal is more useful in aquatic diets than in chicken diets, where they can make the chickens taste like fish. Unfortunately, this substitution can result in lower levels of the highly valued omega-3 content in the farmed product. However, when vegetable oil is used in the growing diet as an energy source and a different finishing diet containing high omega-3 content fatty acids from either fish oil, algae oils, or some vegetable oils are used a few months before harvest, this problem is eliminated.

As of 2008, 50-80% of the world fish oil production is fed to farmed salmonids.

Farm raised salmonids are also fed the carotenoids astaxanthin and canthaxanthin, so their flesh colour matches wild salmon, which also contain the same carotenoid pigments from their diet in the wild.

On a dry-dry basis, 2–4 kg of wild-caught fish are needed to produce 1 kg of salmon. The ratio may be reduced if non-fish sources are added. Wild salmon require about 10 kg of forage fish to produce 1 kg of salmon, as part of the normal trophic level energy transfer. The difference between the two numbers is related to farmed salmon feed containing other ingredients beyond fish meal and because farmed fish do not expend energy hunting.

In 2017 it was reported that the American company Cargill has been researching with EWOS on alternative feeds in its RAPID feed and COMPASS programs in Norway. These methods studied macronutrient profiles of fish feed based upon geography and season. Using RAPID feed, salmon farms reduced the time to maturity of salmon to about 15 months, in a period one-fifth faster than usual.

Harvesting

Modern harvesting methods are shifting towards using wet-well ships to transport live salmon to the processing plant. This allows the fish to be killed, bled, and filleted before rigor has occurred. This results in superior product quality to the customer, along with more humane processing. To obtain maximum quality, minimizing the level of stress is necessary in the live salmon until actually being electrically and percussively killed and the gills slit for bleeding. These improvements in processing time and freshness to the final customer are commercially significant and forcing the commercial wild fisheries to upgrade their processing to the benefit of all seafood consumers.

An older method of harvesting is to use a sweep net, which operates a bit like a purse seine net. The sweep net is a big net with weights along the bottom edge. It is stretched across the pen with the bottom edge extending to the bottom of the pen. Lines attached to the bottom corners are raised, herding some fish into the purse, where they are netted. Before killing, the fish are usually rendered unconscious in water saturated in carbon dioxide, although this practice is being phased out in some countries due to ethical and product quality concerns. More advanced systems use a percussive-stun harvest system that kills the fish instantly and humanely with a blow to the head from a pneumatic piston. They are then bled by cutting the gill arches and immediately immersing them in iced water. Harvesting and killing methods are designed to minimize scale loss, and avoid the fish releasing stress hormones, which negatively affect flesh quality.

Wild versus Farmed

Wild salmonids are captured from wild habitats using commercial fishing techniques. Most wild salmonids are caught in North American, Japanese, and Russian fisheries. The following table shows the changes in production of wild salmonids and farmed salmonids over a period of 25 years, as reported by the FAO. Russia, Japan and Alaska all operate major hatchery based stock enhancement programs that are really ocean ranching. The resulting fish hatchery fish are defined as "wild" for FAO and marketing purposes.

Salmonid production in tonnes by species					
	1982		2007		2013
Species	Wild	Farmed	Wild	Farmed	
Atlantic salmon	10,326	13,265	2,989	1,433,708	2,087,110
Steelhead		171,946		604,695	
Coho salmon	42,281	2,921	17,200	115,376	
Chinook salmon	25,147		8,906	11,542	
Pink salmon	170,373		495,986		
Chum salmon	182,561		303,205		
Sockeye salmon	128,176		164,222		

Total salmonid production				
	1982		2007	
	tonnes	percent	tonnes	percent
Wild	558,864	75	992,508	31
Farmed	188,132	25	2,165,321	69
Overall	746,996		3,157,831	

Issues

The US in their dietary guidelines for 2010 recommends eating 8 ounces per week of a variety of seafood and 12 ounces for lactating mothers, with no upper limits set and no restrictions on eating farmed or wild salmon. In 2018, Canadian dietary guidelines recommended eating at least two servings of fish each week and choosing fish such as char, herring, mackerel, salmon, sardines, and trout.

Currently, much controversy exists about the ecological and health impacts of intensive salmonid aquaculture. Of particular concern are the impacts on wild salmonids and other marine life and on the incomes of commercial salmonid fishermen. However, the 'enhanced' production of salmon juveniles – which for instance lead to a double-digit proportion (20-50%) of the Alaska's yearly 'wild' salmon harvest - is not void of controversy, and the Alaska salmon harvest are highly dependent on the operation of Alaska's Regional Aquaculture Associations. Furthermore, the sustainability of enhanced/hatchery-based 'wild' caught salmon has long been hotly debated, both from a scientific and political/marketing perspective. Such debate and positions were central to a 'halt' in the re-certification of Alaska salmon fisheries by the Marine Stewardship Council (MSC) in 2012. The Alaska salmon fisheries subsequently re-attained MSC-certification status; however the heavily hatchery-dependent Prince William Sound (PWS) unit of certification ("one of the most valuable fishing area in the State") was for several years excluded from the MSC-certification (it remained 'under assessment' pending further analysis).

Disease and Parasites

In 1972, *Gyrodactylus*, a monogenean parasite, was introduced with live trout and salmon from Sweden (Baltic stocks are resistant to it) into government-operated hatcheries in Norway. From the hatcheries, infected eggs, smolt, and fry were implanted in many rivers with the goal to strengthen the wild salmon stocks, but caused instead devastation to some of the wild salmon populations affected.

In 1984, infectious salmon anemia (ISAv) was discovered in Norway in an Atlantic salmon hatchery. Eighty percent of the fish in the outbreak died. ISAv, a viral disease, is now a major threat to the viability of Atlantic salmon farming. It is now the first of the diseases classified on List One of the European Commission's fish health regimen. Amongst other measures, this requires the total eradication of the entire fish stock should an outbreak of the disease be confirmed on any farm. ISAv seriously affects salmon farms in Chile, Norway, Scotland, and Canada, causing major economic losses to infected farms. As the name implies, it causes severe anemia of infected fish. Unlike mammals, the red blood cells of fish have DNA, and can become infected with viruses. The fish develop pale gills, and may swim close to the water surface, gulping for air. However, the disease can also develop without the fish showing any external signs of illness, the fish maintain a normal appetite, and then they suddenly die. The disease can progress slowly throughout an infected farm, and in the worst cases, death rates may approach 100%. It is also a threat to the dwindling stocks of wild salmon. Management strategies include developing a vaccine and improving genetic resistance to the disease.

In the wild, diseases and parasites are normally at low levels, and kept in check by natural predation on weakened individuals. In crowded net pens, they can become epidemics. Diseases and parasites also transfer from farmed to wild salmon populations. The European Commission concluded, "The reduction of wild salmonid abundance is also linked to other factors but there is more and more scientific evidence establishing a direct link between the number of lice-infested wild fish and the presence of cages in the same estuary." It is reported that wild salmon on the west coast of Canada are being driven to extinction by sea lice from nearby salmon farms. These predictions have been disputed by other scientists and recent harvests have indicated that the predictions were in error. In 2011, Scottish salmon farming introduced the use of farmed wrasse for the purpose of cleaning farmed salmon of ectoparasites.

Globally, salmon production fell around 9% in 2015, in large part due to acute outbreaks of sea lice in Scotland and Norway. Lasers are used to reduce lice infections.

Pollution and Contaminants

Salmonid farms are typically sited in marine ecosystems with good water quality, high water exchange rates, current speeds fast enough to prevent pollution of the bottom but slow enough to prevent pen damage, protection from major storms, reasonable water depth, and a reasonable distance from major infrastructure such as ports, processing plants, and logistical facilities such as airports. Logistical considerations are significant, and feed and maintenance labor must be transported to the facility and the product returned. Siting decisions are complicated by complex, politically driven permit problems in many countries that prevents optimal locations for the farms.

In sites without adequate currents, heavy metals can accumulate on the benthos (seafloor) near the salmon farms, particularly copper and zinc.

Contaminants are commonly found in the flesh of farmed and wild salmon. Health Canada in 2002 published measurements of PCBs, dioxins and furans and PDBEs in several varieties of fish. The farmed salmonids population had nearly 3 times the level of PCBs seen in the wild population. The farmed salmonids population had more than 3 times the level of PDBEs seen in the wild population. The farmed salmonids population had nearly twice the level of dioxins and furans seen in the wild population. On the other hand, "Update of the monitoring of levels of dioxins and PCBs in food and feed", a 2012 study from the European Food Safety Authority, stated that farmed salmon and trout contained on average a many times lesser fraction of dioxins and PCBs than wild-caught salmon and trout."

Reported in *Science*, analysed farmed and wild salmon for organochlorine contaminants. They found the contaminants were higher in farmed salmon. Within the farmed salmon, European (particularly Scottish) salmon had the highest levels, and Chilean salmon the lowest. The FDA and Health Canada have established a tolerance/limit for PCBs in commercial fish of 2000 ppb A follow up study confirmed this, and found levels of dioxins, chlorinated pesticides, PCBs and other contaminants up to ten times greater in farmed salmon than wild Pacific salmon. On a positive note, further research using the same fish samples used in the previous study, showed that farmed salmon contained levels of beneficial fatty acids that were two to three times higher than wild salmon. A follow-up benefit-risk analysis on salmon consumption balanced the cancer risks with the (n−3) fatty acid advantages of salmon consumption. For this reason, current methods for this type of analysis take into consideration the lipid content of the sample in question. PCBs specifically are lipophilic, so are found in higher concentrations in fattier fish in general, thus the higher level of PCB in the farmed fish is in relation to the higher content of beneficial n−3 and n−6 lipids they contain. They found that recommended levels of (n-3) fatty acid consumption can be achieved eating farmed salmon with acceptable carcinogenic risks, but recommended levels of (n-3) EPA+DHA intake cannot be achieved solely from farmed (or wild) salmon without unacceptable carcinogenic risks. The conclusions of this paper from 2005 were that:

> "Consumers should not eat farmed fish from Scotland, Norway and eastern Canada more than three times a year; farmed fish from Maine, western Canada and Washington state no more than three to six times a year; and farmed fish from Chile no more than about six times a year. Wild chum salmon can be consumed safely as often as once a week, pink salmon, Sockeye and Coho about twice a month and Chinook just under once a month."

Russia banned importing chilled fish from Norway, after samples of Norwegian farmed fish showed high levels of heavy metals. According to the Russian Minister of Agriculture Aleksey Gordeyev, levels of lead in the fish were 10 to 18 times higher than Russian safety standards and cadmium levels were almost four times higher.

Pollutants or Toxins Introduced by Pisciculturists

Eight Norwegian salmon producers were caught in unauthorized and unlabeled use of nitrite in smoked and cured salmon. Norway applies EU regulations on food additives, according to which nitrite is allowed as a food additive in certain types of meat, but not fish. Fresh salmon was not affected.

Kurt Oddekalv, leader of the Green Warriors of Norway, argues that the scale of fish farming in Norway is unsustainable. Huge volumes of uneaten feed and fish excrement pollute the seabed, while chemicals designed to fight sea lice find their way into the food chain. He says: "If people knew this, they wouldn't eat salmon", describing the farmed fish as "the most toxic food in the world". Don Staniford—the former scientist turned activist/investigator and head of a small Global Alliance Against Industrial Aquaculture—agrees, saying that a 10-fold increase in the use of some chemicals was seen in the 2016-2017 timeframe. The use of the toxic drug emamectin is rising fast. The levels of chemicals used to kill sea lice have breached environmental safety limits more than 100 times in the last 10 years.

Impact on Wild Salmonids

Farmed salmonids can, and often do, escape from sea cages. If the farmed salmonid is not native, it can compete with native wild species for food and habitat. If the farmed salmonid is native, it can interbreed with the wild native salmonids. Such interbreeding can reduce genetic diversity, disease resistance, and adaptability. About 500,000 salmon and trout escaped from ocean net pens off Norway. Around Scotland, 600,000 salmon were released during storms. Commercial fishermen targeting wild salmon frequently catch escaped farm salmon. At one stage, in the Faroe Islands, 20 to 40 percent of all fish caught were escaped farm salmon. About 263,000 farmed non-native Atlantic salmon escaped from a net in Washington waters in the 2017 Cypress Island Atlantic salmon pen break.

Sea lice, particularly *Lepeophtheirus salmonis* and various *Caligus* species, including *C. clemensi* and *C. rogercresseyi*, can cause deadly infestations of both farm-grown and wild salmon. Sea lice are naturally occurring and abundant ectoparasites which feed on mucus, blood, and skin, and migrate and latch onto the skin of salmon during planktonic nauplii and copepodid larval stages, which can persist for several days. Large numbers of highly populated, open-net salmon farms can create exceptionally large concentrations of sea lice; when exposed in river estuaries containing large numbers of open-net farms, many young wild salmon are infected, and do not survive as a result. Adult salmon may survive otherwise critical numbers of sea lice, but small, thin-skinned juvenile salmon migrating to sea are highly vulnerable. In 2007, mathematical studies of data available from the Pacific coast of Canada indicated the louse-induced mortality of pink salmon in some regions was over 80%. Later that year, in reaction to the 2007 mathematical study mentioned above, Canadian federal fisheries scientists Kenneth Brooks and Simon Jones published a critique titled “Perspectives on Pink Salmon and Sea Lice: Scientific Evidence Fails to Support the Extinction Hypothesis “ The time since these studies has shown a general increase in abundance of Pink Salmon in the Broughton Archipelago. Another comment in the scientific literature by Canadian Government Fisheries scientists Brian Riddell and Richard Beamish et al. came to the conclusion that there is no correlation between farmed salmon louse numbers and returns of pink salmon to the Broughton Archipelago. And in relation to the 2007 Krkosek extinction theory: “the data was [sic] used selectively and conclusions do not match with recent observations of returning salmon”.

A 2008 meta-analysis of available data shows that salmonid farming reduces the survival of associated wild salmonid populations. This relationship has been shown to hold for Atlantic, steelhead, pink, chum, and coho salmon. The decrease in survival or abundance often exceeds 50%. However, these studies are all correlation analysis and correlation doesn’t equal causation, especially when similar salmon declines were occurring in Oregon and California, which have no salmon aquaculture or marine net pens. Independent of the predictions of the failure of salmon runs in Canada indicated by these studies, the wild salmon run in 2010 was a record harvest.

A 2010 study that made the first use of sea lice count and fish production data from all salmon farms on the Broughton Archipelago found no correlation between the farm lice counts and wild salmon survival. The authors conclude that the 2002 stock collapse was not caused by the farm sea lice population: although the farm sea lice population during the out-migration of juvenile pink salmon was greater in 2000 than that of 2001, there was a record salmon returning to spawn in 2001 compared with a 97% collapse in 2002. The authors also note that initial studies had not investigated bacterial and viral causes for the event despite reports of bleeding at the base of the fins, a symptom often associated with infections, but not with sea lice exposure under laboratory conditions.

Wild salmon are anadromous. They spawn inland in fresh water and when young migrate to the ocean where they grow up. Most salmon return to the river where they were born, although some stray to other rivers. Concern exists about of the role of genetic diversity within salmon runs. The resilience of the population depends on some fish being able to survive environmental shocks, such as unusual temperature extremes. The effect of hatchery production on the genetic diversity of salmon is also unclear.

Genetic Modification

Salmon have been genetically modified in laboratories so they can grow faster. A company, Aqua Bounty Farms, has developed a modified Atlantic salmon which grows nearly twice as fast (yielding a fully grown fish at 16–18 months rather than 30), and is more disease resistant, and cold tolerant. It also requires 10% less food. This was achieved using a chinook salmon gene sequence affecting growth hormones, and a promoter sequence from the ocean pout affecting antifreeze production. Normally, salmon produce growth hormones only in the presence of light. The modified salmon does not switch growth hormone production off. The company first submitted the salmon for FDA approval in 1996. In 2015, FDA has approved the AquAdvantage Salmon for commercial production. A concern with transgenic salmon is what might happen if they escape into the wild. One study, in a laboratory setting, found that modified salmon mixed with their wild cohorts were aggressive in competing, but ultimately failed.

Impact on Wild Predatory Species

Sea cages can attract a variety of wild predators which can sometimes become entangled in associated netting, leading to injury or death. In Tasmania, Australian salmon-farming sea cages have entangled white-bellied sea eagles. This has prompted one company, Huon Aquaculture, to sponsor a bird rehabilitation centre and try more robust netting.

Impact on Forage Fish

The use of forage fish for fish meal production has been almost a constant for the last 30 years and at the maximum sustainable yield, while the market for fish meal has shifted from chicken, pig, and pet food to aquaculture diets. This market shift at constant production appears an economic decision implying that the development of salmon aquaculture had no impact on forage fish harvest rates.

Fish do not actually produce omega-3 fatty acids, but instead accumulate them from either consuming microalgae that produce these fatty acids, as is the case with forage fish like herring and sardines, or consuming forage fish, as is the case with fatty predatory fish like salmon. To satisfy this requirement, more than 50% of the world fish oil production is fed to farmed salmon.

In addition, salmon require nutritional intakes of protein, which is often supplied in the form of fish meal as the lowest-cost alternative. Consequently, farmed salmon consume more fish than they generate as a final product, though considerably more preferred as food.

Hatch and Release

Another form of salmon production, which is safer but less controllable, is to raise salmon in hatcheries until they are old enough to become independent. They are then released into rivers, often in an attempt to increase the salmon population. This practice was very common in countries such as Sweden before the Norwegians developed salmon farming, but is seldom done by private companies, as anyone may catch the salmon when they return to spawn, limiting a company's chances of benefiting financially from their investment. Because of this, the method has mainly been used by various public authorities and nonprofit groups, such as the Cook Inlet Aquaculture Association, as a way of artificially increasing salmon populations in situations where they have declined due to overharvest, construction of dams, and habitat destruction or disruption. Unfortunately, negative consequences to this sort of population manipulation can occur, including genetic "dilution" of the wild stocks, and many jurisdictions are now beginning to discourage supplemental fish planting in favour of harvest controls and habitat improvement and protection. A variant method of fish stocking, called ocean ranching, is under development in Alaska. There, the young salmon are released into the ocean far from any wild salmon streams. When it is time for them to spawn, they return to where they were released where fishermen can then catch them.

Land-raised Salmon

Recirculating acquaculture systems make it possible to farm salmon entirely on land, which as of 2019 is an ongoing initiative in the industry. However, large farmed salmon companies such as Mowi and Cermaq were not investing in such systems. In the United States, a major investor in the effort was Atlantic Sapphire, which planned to bring salmon raised in Florida to market in 2021. Other companies investing in the effort include Nordic Acquafarms and Whole Oceans.

Species

Atlantic Salmon

In their natal streams, Atlantic salmon are considered a prized recreational fish, pursued by avid fly anglers during its annual runs. At one time, the species supported an important commercial fishery and a supplemental food fishery. However, the wild Atlantic salmon fishery is commercially dead; after extensive habitat damage and overfishing, wild fish make up only 0.5% of the Atlantic salmon available in world fish markets. The rest are farmed, predominantly from aquaculture in Chile, Canada, Norway, Russia, the United Kingdom, and Tasmania.

Atlantic salmon

Atlantic salmon is, by far, the species most often chosen for farming. It is easy to handle, grows well in sea cages, commands a high market value, and adapts well to being farmed away from its native habitats.

Adult male and female fish are anesthetized. Eggs and sperm are "stripped", after the fish are cleaned and cloth dried. Sperm and eggs are mixed, washed, and placed into fresh water. Adults recover in flowing, clean, well-aerated water. Some researchers have studied cryopreservation of the eggs.

Fry are generally reared in large freshwater tanks for 12 to 20 months. Once the fish have reached the smolt phase, they are taken out to sea, where they are held for up to two years. During this time, the fish grow and mature in large cages off the coasts of Canada, the United States, or parts of Europe. Generally, cages are made of two nets; inner nets, which wrap around the cages, hold the salmon while outer nets, which are held by floats, keep predators out.

Many Atlantic salmon escape from cages at sea. Those salmon that further breed tend to lessen the genetic diversity of the species leading to lower survival rates, and lower catch rates. On the West Coast of North America, the non-native salmon could be an invasive threat, especially in Alaska and parts of Canada. This could cause them to compete with native salmon for resources. Extensive efforts are underway to prevent escapes and the potential spread of Atlantic salmon in the Pacific and elsewhere. The risk of Atlantic Salmon becoming a legitimate invasive threat on the Pacific Coast of N. America is questionable in light of both Canadian and American governments deliberately introducing this species by the millions for a 100-year period starting in the 1900s. Despite these deliberate attempts to establish this species on the Pacific coast; no established populations have been reported.

In 2007, 1,433,708 tonnes of Atlantic salmon were harvested worldwide with a value of $7.58 billion. Ten years later, in 2017, over 2 million tonnes of farmed Atlantic salmon were harvested.

Steelhead

Rainbow trout

Male ocean phase steelhead salmon

In 1989, steelhead were reclassified into the Pacific trout as *Oncorhynchus mykiss* from the former binominals of *Salmo gairdneri* (Columbia River redband trout) and *S. irideus* (coastal rainbow trout). Steelhead are an anadromous form of rainbow trout that migrate between lakes and rivers and the ocean, and are also known as steelhead salmon or ocean trout.

Steelhead are raised in many countries throughout the world. Since the 1950s, production has grown exponentially, particularly in Europe and recently in Chile. Worldwide, in 2007, 604,695 tonnes of farmed steelhead were harvested, with a value of $2.59 billion. The largest producer is Chile. In Chile and Norway, the ocean-cage production of steelhead has expanded to supply export markets. Inland production of rainbow trout to supply domestic markets has increased strongly in countries such as Italy, France, Germany, Denmark, and Spain. Other significant producing countries include the United States, Iran, Germany, and the UK. Rainbow trout, including juvenile steelhead in fresh water, routinely feed on larval, pupal, and adult forms of aquatic insects (typically caddisflies, stoneflies, mayflies, and aquatic dipterana). They also eat fish eggs and adult forms of terrestrial insects (typically ants, beetles, grasshoppers, and crickets) that fall into the water. Other prey include small fish up to one-third of their length, crayfish, shrimp, and other crustaceans. As rainbow trout grow, the proportion of fish consumed increases in most populations. Some lake-dwelling forms may become planktonic feeders. In rivers and streams populated with other salmonid species, rainbow trout eat varied fish eggs, including those of salmon, brown and cutthroat trout, mountain whitefish, and the eggs of other rainbow trout. Rainbows also consume decomposing flesh from carcasses of other fish. Adult steelhead in the ocean feed primarily on other fish, squid, and amphipods. Cultured steelhead are fed a diet formulated to closely resemble their natural diet that includes fish meal, fish oil, vitamins and minerals, and the carotenoid asthaxanthin for pigmentation.

The steelhead is especially susceptible to enteric redmouth disease. Considerable research has been conducted on redmouth disease, as its implications for steelhead farmers are significant. The disease does not affect humans.

Coho Salmon

Male ocean phase Coho salmon

The Coho salmon is the state animal of Chiba, Japan. Coho salmon mature after only one year in the sea, so two separate broodstocks (spawners) are needed, alternating each year. Broodfish are selected from the salmon in the seasites and "transferred to freshwater tanks for maturation and spawning".

Worldwide, in 2007, 115,376 tonnes of farmed Coho salmon were harvested with a value of $456 million. Chile, with about 90 percent of world production, is the primary producer with Japan and Canada producing the rest.

Chinook Salmon

Male ocean-phase Chinook

Male freshwater-phase Chinook

Chinook salmon are the state fish of Oregon, and are known as "king salmon" because of their large size and flavourful flesh. Those from the Copper River in Alaska are particularly known for their color, rich flavor, firm texture, and high omega-3 oil content. Alaska has a long-standing ban on finfish aquaculture that was enacted in 1989.

Worldwide, in 2007, 11,542 tonnes (1,817,600 st) of farmed Chinook salmon were harvested with a value of $83 million. New Zealand is the largest producer of farmed king salmon, accounting for over half of world production. Most of the salmon are farmed in the sea (mariculture) using a method sometimes called sea-cage ranching, which takes place in large floating net cages, about 25 m across and 15 m deep, moored to the sea floor in clean, fast-flowing coastal waters. Smolt (young fish) from freshwater hatcheries are transferred to cages containing several thousand salmon, and remain there for the rest of their lives. They are fed fishmeal pellets high in protein and oil.

Chinook salmon are also farmed in net cages placed in freshwater rivers or raceways, using techniques similar to those used for sea-farmed salmon. A unique form of freshwater salmon farming occurs in some hydroelectric canals in New Zealand. A site in Tekapo, fed by fast, cold waters from the Southern Alps, is the highest salmon farm in the world, 677 m (2,221 ft) above sea level.

Before they are killed, cage salmon are sometimes anaesthetised with a herbal extract. They are then spiked in the brain. The heart beats for a time as the animal is bled from its sliced gills. This method of relaxing the salmon when it is killed produces firm, long-keeping flesh. Lack of disease in wild populations and low stocking densities used in the cages means that New Zealand salmon farmers do not use antibiotics and chemicals that are often needed elsewhere.

Aquaculture of Tilapia

Aquaculture of tilapias provides a classic example of a success story of a species group outside its natural range of distribution. The group currently contributes about 3.8 percent to the cultured fish and shellfish production of about 40 million tonnes globally. The current aquaculture production of tilapias is about 1.5 million tonnes, the great bulk of which takes place in Asia accounting for nearly 80 percent of the total world production. It is important to note, however, that tilapia culture in Africa and South America is also increasing.

Prior to the mid-1990s, the yield of tilapia from capture fisheries was greater than that from aquaculture. Currently, the latter accounts for approximately 2.5 times the production from capture fisheries. Tilapia aquaculture production increased from 28 000 tonnes to 1.504 million tonnes globally from 1970 to 2002; in Asia and the Pacific, production increased from 23 000 tonnes to 1.192 million tonnes equivalent to an annual growth rate of 13.2 percent and 13.1 percent, respectively. In contrast, capture fisheries for tilapias have grown at the rate of 3.5 percent per annum.

The growth of tilapia culture can be exemplified by comparing production with cyprinids and salmonids in different time periods. The trend shows that in all three groups of fish the annual rate of increase declined with time, as in the case of the aquaculture industry as a whole. However, the rate of increase in tilapia culture exceeded that of cyprinids and salmonids over all the time periods considered, with tilapia culture recording the highest rate of annual growth over the last two decades among all finfish groups. The growth of tilapia culture in Asia and the Pacific however, lagged slightly behind to that of the world. For example, the percent annual increase in tilapia culture in Asia and the Pacific for the 20 and 5 year periods was 12.5 (vs. 12.6), 10.9 (vs. 11.9) and 7.7 (vs. 10.0), respectively. This lower rate of growth in Asia and the Pacific is more a reflection of increased tilapia production in countries such as Egypt, where O. niloticus production increased from 9 000 tonnes in 1980 to 168 000 tonnes in 2002. Also the rate of growth observed, in all three groups of fish is considerably higher than that witnessed in the sector as a whole. Although a number of tilapia species has been introduced into the region, only a small number of these are cultured.

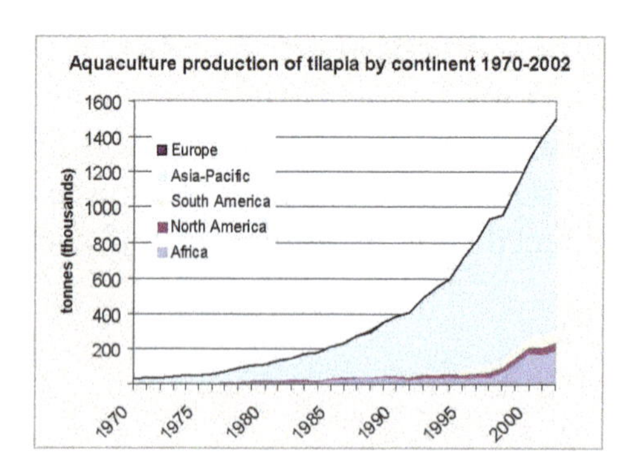

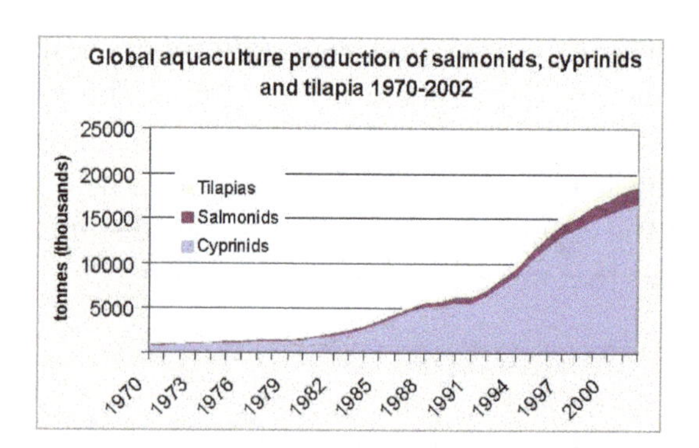

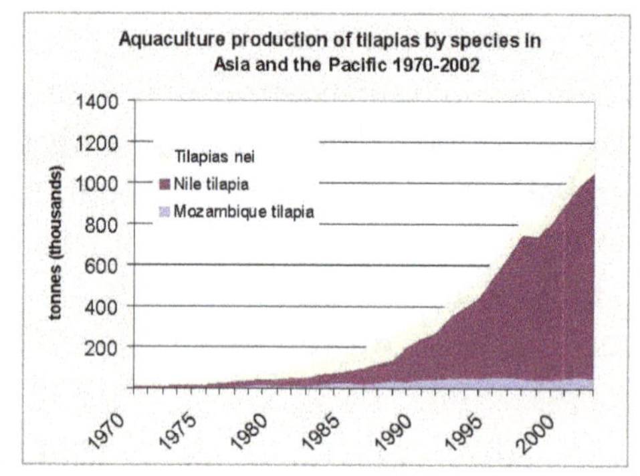

Oreochromis mossambicus and O. niloticus are the most widely cultured tilapias in the world. In 15 countries in Asia and the Pacific, only four tilapia species are cultured, dominated by O. mossambicus (five countries) and O. niloticus (10 countries); the latter accounts for more than 90 percent of the production and its contribution to aquaculture production has been increasing steadily.

Since 1995, some South American nations have become significant contributors to cultured tilapia production. Beginning in 1989, Peoples Republic of China (PR China) has dominated global tilapia aquaculture production, accounting for 47 percent of global production in 2002. In Egypt, cultured tilapia production increased from 9 000 tonnes in 1980 to 168 000 tonnes in 2002, thus attaining the second highest production in global tilapia culture.

Farmed red tilapia in Sri Lanka

The principal nations in Asia and the Pacific which have adopted tilapia culture are PR

China, Indonesia, the Philippines, Thailand and Taiwan Province of China. Changes in O. niloticus production in these countries indicate that PR China currently accounts for over 70 percent of the region's production, an increase from 39 percent of Asia and the Pacific production in 1988, before the reported rapid increase in Nile tilapia culture in PR China. This has led to a decreased share of production from the other countries of the region, even when production has exhibited growth. For example, in the Philippines, the proportional contribution to regional Nile tilapia production was 27% in 1988 and only 10% in 2002, even though production increased from 27 000 tonnes to 104 000 tonnes over the period. Indonesia continues to dominate O. mossambicus culture, accounting for nearly 50 000 tonnes and more than 90 percent of the global total in 2002. Over 60 percent of this production is reported as originating in brackish waters.

One of the significant changes that has occurred over the last decade is the increase in tilapia culture in brackish waters. For 2002, brackishwater culture production was 193 000 tonnes. This represents a tripling from the relatively constant values (approximately 60 000 tonnes) for the period 1987-1997. O. niloticus accounts for about 73 percent of the brackishwater production, with most of the increase driven by large increases in production in Egypt. On the other hand, brackishwater culture of O. mossambicus has not increased as dramatically over the last decade. Egypt dominates O. niloticus brackishwater culture with 138 000 tonnes produced in 2002.

Table: The number of countries/territories in each continent where different species of tilapias are cultured. Many countries do not report to species level so Orechromis spp. is included as well.

Continent/Region		Oreochromis spp.					Tilapia spp.	
	not specified (spp.)	niloticus	mossambicus	aureus	spilurus	andersonii	rendalli	zillii
N. America	4	7	2	3	0	0	0	0
S. America	3	4	1	0	0	0	0	0
Africa	4	13	1	0	0	1	0	1
Asia-Pacific	9	10	5	0	1	0	0	0
Europe	1	0	0	0	0	0	0	0

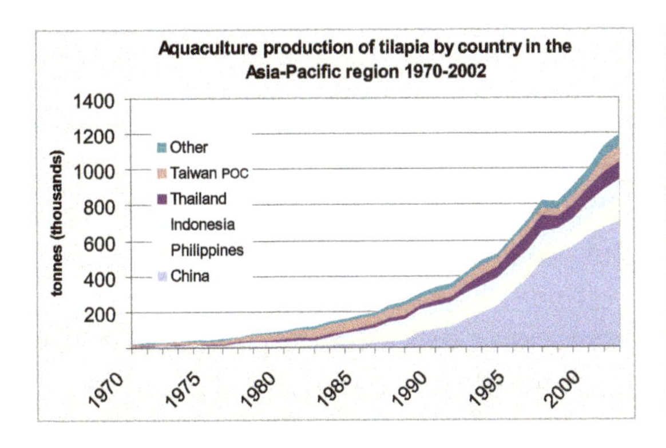

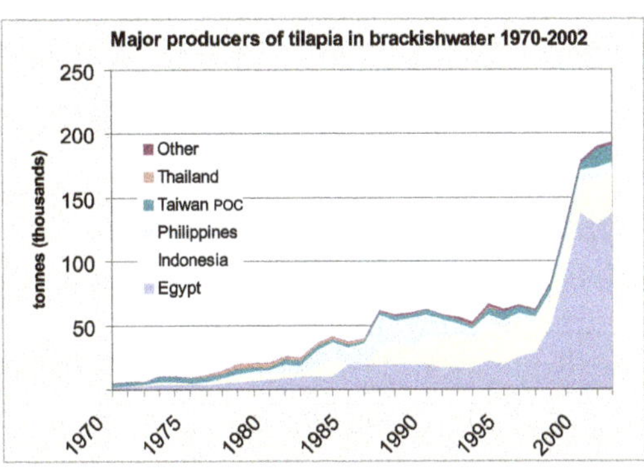

Culture Practices

Culture practices of tilapias in the world are very diverse, perhaps the most diverse among all aquaculture species in the world. It is a group of fish that could be cultured at many desired intensities, thus, appealing to all socio-economic strata, enabling the culture practices to be adjusted to suit their economic capabilities. Oreochromis niloticus is commonly cultured in backyard and home garden ponds to supplement the income of poor households as well as provide a fresh source of animal proteins to the family. In such situations, the cultured stock is often fed with kitchen waste and supplemented by relatively readily available, often low cost agricultural by-products such as rice bran. However, the direct nutritional value of the latter to the stock is not known and in all probability rather low; the inputs act more as a fertilizer. Oreochromis niloticus is cultured in relatively poor quality waters, including: (a) sewage fed ponds (e.g. commercial culture in Calcutta, India); and (b) primary and secondary treated waste effluents (e.g. Egypt). So far, there have not been any reports of detrimental effects of consumption of fish reared in sewage-fed farms on human health even as the practice has been in operation since the 1930s.

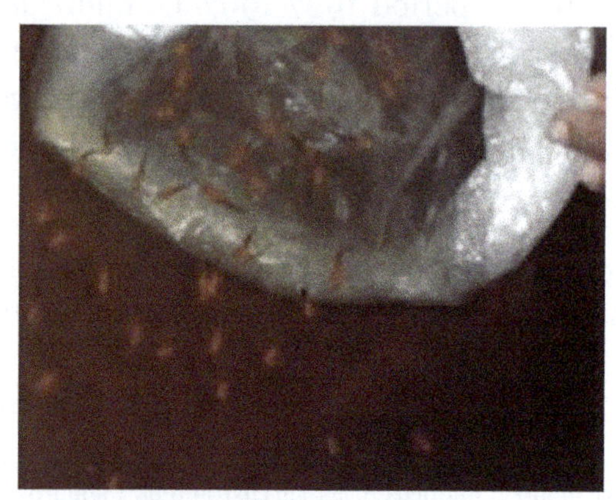

Releasing red tilapia fry, Sri Lanka

Tilapias are generally used in other aquaculture systems. Oreochromis niloticus is often used in integrated-fish culture. The most extreme development in this system is the integration of commercial poultry farming with O. niloticus culture. Cultured stocks are not fed, but depend on poultry waste and the natural food production in eutrophic ponds. Oreochromis niloticus is also being increasingly used in rice-fish culture, a traditional system which is gaining revival in Asia and is reported to enhance the overall yields in practices in PR China and Bangladesh. Similarly, the potential of using O. niloticus in biogas slurry has been demonstrated with a view to integrating biogas technology in food production.

Tilapia pond culture practices are, in almost all cases, conducted in static systems, which require water replenishment between grow-out cycles and aeration is rarely used. Such minimal requirements reduce the initial investment and also keep the recurrent cost relatively low and therefore, more affordable for poorer sectors of the community. Cage culture of tilapias is also relatively popular, the intensity and the sophistication varying from practice to practice and requiring minimal investment. Tilapia cage culture is effective as a means of providing alternative livelihood to displaced persons during reservoir impoundment (e.g. Jatilnuhur reservoir in Indonesia and Batan Ai reservoir in Malaysia). Operations in these cases are done on an industrial scale when cage construction, feed supplies, marketing strategies etc. are provided as an integrated package.

Most interestingly, tilapia culture is also being conducted in recirculation or closed cycle systems in temperate countries such as Canada, which incorporates expertise on greenhouse operations. This is an evidence of a growing demand for cultured tilapia in many countries and the subsequent expansion of tilapia culture practices throughout the world.

The diversity of tilapia culture practices is also borne by the fact that tilapias are high salinity tolerant species and could be cultured in brackish waters and in sea cages. For sea cage culture, the appropriate species is O. spilurus. As mentioned earlier, tilapia production in brackish waters have grown steadily over the last decade with Egypt being the largest producer.

The returns from tilapia culture differ according to the environment, system and practice of culture. According to Dey, the best tilapia culture practices in Asia and the Pacific are encountered in Taiwan Province of China yielding 12 to 17 tonnes/ ha/yr in ponds, followed by mean yields of 6.6, 6.3, 3.0 and 1.7 tonnes/ha/yr in Thailand, the Philippines, Viet Nam and Bangladesh, respectively. Cage culture operations in the Philippines and in PR China on the other hand, yielded 0.5 and 5.5 tonnes/100 m2, respectively. Similarly, the mean size at harvest varied considerably among countries; between culture systems, the highest (460 g) being in cage cultured O. niloticus in PR China.

Why have tilapias been successful as a cultured species group throughout the tropical countries? Bearing in mind that the bulk of tilapia production is a result of farming , its success can be primarily attributed to the following:

Cage culture of tilapia in Batanai reservoir, Sarawak, East Malaysia

- Relative ease of culture under extensive, semi- and intensive practices, thus relatively less limited by the economic status of the farmer compared to most other finfish species;

- Relevant species exhibit many of the desirable traits expected of a species suitable for culture (e.g. relatively high growth rate, wide range of tolerance to physico-chemical characteristics, resistance to disease, easiness of propagation, etc.);
- Moderately high dress-weight ratio;
- Long shelf-life and;
- As a white fish, tilapia is mild and lends itself to industrial preparations better than most other white fish.

In addition to all of the above, most of the commonly cultured tilapias are easily weaned on to artificial feeds. The group has the ability to derive its nutrition effectively from the natural food in the rearing systems particularly in ponds. This attribute makes it the foremost choice in home-garden and backyard small-scale, subsistence fish culture in developing countries such as Bangladesh, Viet Nam, etc.

Table: Different hybrid combinations of tilapias that have known to produce monosex male progeny.

<table>
<tr><th>Female parent</th><th>Male parent</th><th>Remarks</th></tr>
<tr><td>O. niloticus</td><td></td><td>Applied commercially but results inconsistent</td></tr>
<tr><td>O. niloticus</td><td>O. macrochir</td><td rowspan="2">Majority of broods are all male; some commercial application too</td></tr>
<tr><td>O. niloticus</td><td>O. urolepis hornorum</td></tr>
<tr><td>O. niloticus</td><td>O. variables</td><td rowspan="7">All progenies monosex</td></tr>
<tr><td>O. mossambicus</td><td>O. aureus</td></tr>
<tr><td>O. mossambicus</td><td>O. urolepis hornorum</td></tr>
<tr><td>O. spilurus niger</td><td>O. macrochir</td></tr>
<tr><td>O. spilurus niger</td><td>O. urolepis hornorum</td></tr>
<tr><td>O. aureus</td><td>O. urolepis hornorum</td></tr>
<tr><td>T. zillii</td><td>O. andersonii</td></tr>
</table>

Notable Phases in the Development of Tilapia Culture

Culture of tilapias has gone through a number developmental phases since the nineteenth century. In confined environments such as ponds, early maturation of the species, which resulted to overpopulation and eventually stunting, needs to be addressed. The earliest approach to overpopulation was stocking of predators, and few examples include Ophicephalus spp. and Clarias spp. in Asia, L. niloticus and Micropterus salmoides) in Africa and Cichalosoma managuense in Central America. Such a strategy requires the following: (a) optimization of predator to prey ratios; and (b) appropriate timing of the introduction with the correct number of predators of appropriate sizes. The effectiveness of such a strategy was limited and "fine tuning" is required so that an optimal balance is formed between the fry production of tilapias and the voracity of the predators. This strategy does not eliminate the channelling of food energy for egg and fry production thereby decreasing yields that could have been otherwise attained.

Realizing that male tilapias grow faster, further approaches to solving the problem of overpopulation in culture systems was to develop all-male tilapias. Beardmore et al. summarized the potential advantages of the use of monosex animals in aquaculture:

- Higher growth rate;
- Elimination of reproduction;
- Reduction of sexual/territorial behaviour;
- Narrower size range at harvest; and
- Reduction of environmental impacts resulting from escapees.

In view of the above considerations, the initial approach was to produce monosex tilapias through hybridization. Female O. mossambicus x male O. hornorum cross generated nearly all-male hybrids. Other cross selection followed which resulted in all-male and nearly all-male hybrids. Thus, hybridization strategy minimized overpopulation in the culture environments and also resulted in higher yields because of the faster rate of growth of the males. However, large-scale hybrid production did not always succeed due to instability in production of all-male hybrids and increasing appearance of females among progenies. Only few species of commercial value consistently produced allmale F1 generation. The most notable of these was the O. niloticus x O. aureus cross, which tolerated winter periods in Israel and became a focal point for a fresh wave of distribution of the parent species to other regions. Except in a few cases, hybridization did not become established as a commercial method for allmale tilapia production.

The hybridization phase gave way to hormonal sex-reversal as a method of producing large quantities of all-male seed stock, an approach perceived to be more practical. At the time that hormonal sex reversal was gaining popularity, a "shit" in the species was taking place, where O. niloticus became the preferred species for culture in view of its faster growth rate, atractive coloration and supposedly better taste. Consequently, most of the technological advances on sex-reversal were developed for this species, although the initial discoveries were made on O. mossambicus when Clemens and Inslee succeeded in producing all-males by oral administration of 17α-methyltesterone (male hormone) to early stage fry. Since the original discovery the effectiveness of 17α-methyltesterone, sex-reversal in a number of tilapia species has been demonstrated, e.g. O. aureus, O. niloticus, O. hornorum, Tilapia zillii, O. spilurus, among others. In addition, apart from 17αmethyltesterone, the effectiveness of other chemicals incorporated in food and orally administered for sex-reversal in tilapias has also been demonstrated.

Hormonal sex-reversal is effective in early fry stages only. The hormones are almost always provided with the feed by spraying an alcoholic solution of the hormone on the feed and allowing the alcohol to evaporate. The dosage (generally 40 to 60 ppm) and treatment duration for the different hormones used for tilapias are well documented. It is also important to note that the orally administered masculinizing hormones, in particular 17α-methyltesterone, are eliminated (converted into polar metabolites) within 72 h of administration and that by day 10 only traces remained. Indeed, even in instances when 17α-methyltesterone was used as a growth promoter, the chemical was eliminated from muscle within a very short period of time, thus enablling the continued use of the technique without restrictions on the marketing the final product.

The technology for mass production of sex-reversed juveniles was initially developed by Rothbard et al. and has been extended into many hatcheries in the region. On the other hand, Beardmore et al. pointed out the main disadvantages of the method particularly in relation to practical difficulties in providing a uniform dose to all of the stock.

Following hormonal sex-reversal of tilapias and the greater understanding of sexdetermination systems in a number of key species, it became possible to develop techniques for the production of genetically male tilapia (GMT) initially on O. niloticus. This technique is based on creation of males with two Y chromosomes (supermales) that when mated with other genotypes usually to give all-males and generally only a small percentage of females. Genetically male has also been produced, in this instance, through gynogenesis of XY neofemales.

The technique of producing GMT, particularly in the case of O. niloticus, is well established and is commercially adopted now, achieved through several generations of breeding. Beardmore et al. claimed that the YY/GMT technology in O. niloticus is the only genetic technology adopted by the aquaculture industry for the production of all-males. Mair and Little considered the relative advantages, including commercial viability, of producing genetically all-male tilapia and its, as opposed to hormone induced sex- inversion to be: (a) all genetically male progeny; (b) potential to produce 100 % male progeny; (c) potential that no reproduction will occur in the growout phase; (d) applicability to most fry production systems; (e) not labour intensive after the initial phase; (f) lack of consumer resistance; (g) no centralization of fry production; and (h) comparatively higher growth rate.

Perhaps the best established dissemination programme of the GMT technology is in the Philippines done through an organization named Phil-Fishgen. Phil-Fishgen is also involved in research and evaluation of the technology and has an accreditation scheme for private hatcheries. It is estimated that the 32 accredited hatcheries currently hold about 40 000 broodstock sets (1YY male: 3 normal females). According to Mair et al., the broodstocks held in the accredited hatcheries are capable of producing 50 milion GMT fry per year.

Using on-station and on-farm trials, Mair et al. demonstrated that the yield from GMT was 30–40 percent higher compared to normal mixed-sex tilapia (existing Philippine strains) and the mean grow-out period for GMT was 4.6 months, compared to 4.8 and 5.1 months for GIFT (Genetic Improvement of Farmed Tilapia – see next sectionj) fish and sex-reversed tilapia, respectively. The comparatively lower price of fry and fingerlings of GMT compared to prices of sex–reversed and GIFT fish appears to lure more growers, particularly the newer ones, to GMT farming. However, it must be noted that there is increasing evidence that the sex determining mechanisms in fish in general (tilapias in particular) are very plastic and often influenced by environmental factors such as naturally occurring exogenous steroids, temperature and other physical variables and pollution. The environmental factors are often the main cause for the presence of females observed in GMT fish, which at times are potentially capable of mitigating growth advantage otherwise attributed to GMT fish.

Broodstock

Broodstock, or broodfish, are a group of mature individuals used in aquaculture for breeding purposes. Broodstock can be a population of animals maintained in captivity as a source of replacement for, or enhancement of, seed and fry numbers. These are generally kept in ponds or tanks in which environmental conditions such as photoperiod, temperature and pH are controlled. Such populations often undergo conditioning to ensure maximum fry output. Broodstock can also be sourced from wild populations where they are harvested and held in maturation tanks before their seed is collected for grow-out to market size or the juveniles returned to the sea to supplement natural populations. This method, however, is subject to environmental conditions and can be unreliable seasonally, or annually. Broodstock management can improve seed quality and number through enhanced gonadal development and fecundity.

Management

Broodstock management involves manipulating environmental factors surrounding the broodstock to ensure maximum survival, enhance gonadal development and increase fecundity. Such conditioning is necessary to ensure the sustainability of aquaculture production, and to increase the number and quality of eggs produced and control the timing of maturation and spawning. Management of the technologies for gamete production in captivity is one of the essential step for aquaculture that would ensure the growth to this sector. Unfortunately, most fish when reared in captivity condition, exhibit some degree of reproduction dysfunction. Many species of captive fish are able to reach reproduction maturity in aquaculture conditions and gonadal growth occurs normally. However, some of female species often fail final oocyte maturation stage. Hormonal manipulation and acceleration of final oocyte maturation due to the economics of broodstock management is important. For instance, in Salmoniformes, the need to collect the eggs by stripping is a serious limitation, while the time of ovulation must be predicted with accuracy, as over-ripening may take place in minutes or hours after ovulation Therefore, control of broodstock reproductive is essential for the sustainability of commercial aquaculture production.

Choosing species to use requires consideration of the biology of the species. This includes their size at maturity, method of reproduction, feeding behaviour and ability to tolerate adverse conditions Farms also consider whether they grow their own broodstock or obtain them from natural populations. Where natural populations are excluded, the farm can be considered a self-sustaining unit independent of external genetic influence.

Pond-reared broodstocks are selected, often as immature juveniles, and grown out in suitable conditions to sexual maturity. These animals require stable water characteristics and a well-balanced, species-dependent, protein rich diet. This enhances the germinal tissue for future seed stock as it is formed in juveniles.

The pond or tank in which broodfish are held must be a suitable size to hold and condition the broodstock. Dependent on the species involved you need to alter the number of individuals, and often separate the sexes. Sex separation enables the broodstock males and females to be subjected to different conditions where necessary. For example, male and female sturgeons respond to different hormone levels, this also allows more control over eggs and sperm.

The characteristics of the water in which the mature broodstocks are held must be manipulated. The aquaculturist must consider the appropriate oxygen concentration, temperature, and pH of the water all of which can be species specific.

The feeding regime of broodstocks is species specific and requires consideration of timing and composition of the food. Protein, lipid and fatty acid composition is particularly important. The quantity of food intake can be altered to influence spawning and maturity, for example low rations have been shown to reduce the number of fish reaching maturity while increasing the fecundity of those which do.

When fry are desired, spawning can be induced in broodstocks by manipulation of relevant environmental factors. In particular the photoperiod can be altered to imply that it is time to spawn. A shortended photoperiod is known to advance spawning times while a lengthened photoperiod can delay spawning. Artificial light can be used to change the apparent day length and indicate different seasonal features so as to delay spawning. Water temperature can be increased for the same purpose. Following spawning the female broodfish are often stressed and have lost weight. They require extra care and abundant feeding at this time to ensure survival to the next spawning season.

Advantages

Managers can select for reproductive characteristics which influence the egg producing capability of individuals and increase fecundity by providing them with optimal environment and diets. This is further possible in pond-reared populations where traits can be selected for over generations for example, for higher fecundity.

The breeding season and spawning times can be shifted thus expanding the seasonal range of production. This leads to more efficient aquaculture because fry are available to the market year round. Hormonal treatments can advance spawning by two to three weeks. Manipulating photoperiod can alter spawning time by over four months and is cheap and straightforward to achieve.

Broodstock managers can use or select for traits such as fast growth rates or disease resistance over generations to produce more desirable fish. This ability for genetic improvement of stocks is more efficient and produces higher value stock. Broodstocks also enable you to selectively plan and control all matings. Selective breeding is an important part of the domestication of aquaculture species.

Pond-reared broodstocks benefit from the removal of predation which can be a significant cause of mortality in natural populations. They further benefit from the removal of variable environmental impacts.

Holding broodstock in an accessible pond or tank offers readily available breeding adults whenever required.

Disadvantages

When broodstocks are used to supplement natural populations they face different selective pressures to normal. Thus they may not have adequate fitness to survive the natural environment, or can alter and decrease natural genetic diversity due to the bottleneck nature of breeding from a smaller population.

Broodstocks require supplementation from outside sources regularly to prevent negative effects of closed populations. Domestication of broodstocks in hatcheries can reduce reproductive capabilities and alter other genetic characteristics. For example, a trout stock maintained as a closed population for 20 generations showed reduced number and size of egg production.

Penaeidae

Shrimp, particularly of the family Penaeidae, are one of the largest internationally traded species. Native stocks are usually collected as sources of broodstock supply. There are also examples of pond-reared Penaeidae broodstocks. These shrimp are raised in suitable environmental conditions including a 12–14 hour/day photoperiod, a water temperature of 25–29 °C and full seawater salinity with high water exchange rates.

Sydney Rock Oyster

The Sydney rock oyster, *Saccostrea glomerata*, has been farmed in New South Wales, Australia for over 100 years. Due to declines in the supply in the past 30 years, New South Wales introduced a selection program in 1990 to breed faster growing stocks. The utilised broodstocks are held in artificial ponds of around 0.11 ha in size, and at low densities. Broodstocks provided higher numbers of larvae and could be spawned readily providing a more definite source of Sydney rock oysters.

Rainbow Trout

Global production of rainbow trout, *Oncorhynchus mykiss*, requires over 3 billion eggs per year. This number is met because of broodstocks which undergo selection and conditioning in hatcheries. Trout have been reared artificially for over 80 years. Rainbow trout broodstocks are commonly manipulated to delay maturation and spawning time in order to provide eggs regularly and optimise supply. Artificial selection has favoured larger fish due to evidence of correlations between fish size and fecundity.

Commercial Fish Feed

Manufactured feeds are an important part of modern commercial aquaculture, providing the balanced nutrition needed by farmed fish. The feeds, in the form of granules or pellets, provide the nutrition in a stable and concentrated form, enabling the fish to feed efficiently and grow to their full potential.

Many of the fish farmed more intensively around the world today are carnivorous, for example Atlantic salmon, trout, sea bass, and turbot. In the development of modern aquaculture, starting in the 1970s, fishmeal and fish oil were key components of the feeds for these species. They are combined with other ingredients such as vegetable proteins, cereal grains, vitamins and minerals and formed into feed pellets. Wheat, for example, is widely used as it helps to bind the ingredients in the pellets.

Other forms of fish feed being used include feeds made entirely with vegetable materials for species such as carp, moist feeds preferred by some species (easier to make but more difficult to store), and trash fish – that is fish caught and fed directly to larger species being raised in aquaculture pens.

Hatchery Feeds

Specialised feeds are produced for fish hatcheries. In species such as salmon and trout, the newly hatched fry first feed from their yolk sacs and then can be fed with starter feeds. Marine species such as sea bass, sea bream, flounders and turbot consume the nutrition in their yolk sacs during the first few days post hatching and then are fed for several weeks on live prey, in the form of rotifers and brine shrimp (Artemia). Special feeds can be used to enrich the nutritional value of the prey. Rotifers are usually bred in the hatchery while brine shrimp are generally collected from the wild, e.g. salt lakes. Manufactured feed alternatives to brine shrimp are becoming available, offering more consistent nutrition and improved sustainability as demand increases with the growth of aquaculture.

Development of Manufactured Feeds

Until the end of World War II most fish hatcheries relied on raw meat (horse meat in particular) as a dietary staple for trout. In the early 1950s, John E. (Red) Hanson, while working for the New Mexico Game and Fish Department, began experimenting with dietary routine and dry pellet formulations. The first fish feed pellets were introduced to hatchery trout at the Red River Hatchery near Taos. The pellets resulted in improved conversion rates of food intake to fish production, and lead in turn to the wider adoption of fish pellets in hatcheries.

The development of dry pelleted fish feeds to date has two themes. One theme is on improving digestibility and refining the balance of nutrients to match the needs of the different species of fish more precisely at different periods of development. The other theme is to improve the sustainability of the ingredients used. This is being achieved mainly by identifying additional sustainable sources of ingredients, in particular to reduce the need for fishmeal and fish oil. Improving the efficiency of feeding also contributes to sustainability.

Sustainability

Fish feed production in Stokmarknes Norway

Dry fish feed pellets

Traditionally two of the most important ingredients have been fishmeal and fish oil. These come mainly from the processing of fish from the wild catch, usually pelagic species that are generally not suited to processing for human consumption. Fish sold for human consumption attract a higher price than those used to make fishmeal. The fishmeal fisheries are often referred to as reduction fisheries. The world's largest reduction fishery is in the Pacific, off the coast of Peru and Chile and is regulated by the governments of those countries. The North Atlantic is another important source of fish for fishmeal and fish oil. Many major suppliers belong to the International Fishmeal and Fish Oil Organisation.

Fishmeal is a brown, flour-like material made by specialist producers that cook, press, dry and grind the fish. The fish oil is effectively a by-product of this process that proves to be a rich source of energy and fatty acids for fish, including the important long-chain omega-3 fatty acids EPA and DHA now linked to the health benefits associated with eating oily fish such as salmon and mackerel. Fish in general also are good sources of many vitamins and minerals and are often recommended as part of a healthy diet by governmental food agencies.

Because the catches of wild fish must be managed at sustainable levels to ensure the stocks continue to be viable, the available supply of fishmeal and fish oil from these resources will not increase.

The global demand for fish from consumers around the world is increasing. Reasons include the growing population, rising average incomes and greater awareness of fish as part of a healthy diet. The yield from the wild catch cannot be increased sustainably, therefore, in the opinion of observers such as the Food and Agriculture Organization (FAO) of the United Nations, aquaculture must fill the gap. Currently the supply of fish from aquaculture approximately matches that from the wild catch, according to FAO figures.

The current drive in research and development is enabling this to happen by supplementing fishmeal and fish oil with vegetable proteins and oils, while ensuring the fish continue to provide the important health benefits for consumers. Other potential raw material resources are also being explored. For example, the U.S. biotechnology company BioTork is piloting the use of raw materials such as unmarketable papaya and by-products from biodiesel production to produce fish feed components, as well as feeding agricultural waste to algae and fungi that manufacture some of the proteins and omega-3 oils needed for fish food. The US biotechnology company Calysta and the UK/Danish biotech company Unibio opened small plants in the UK and Denmark to produce fish feed from natural gas in 2016.

Modern Fish Feed

Extruded fish feed

Modern fish feeds are made by grinding and mixing together ingredients such as fishmeal, vegetable proteins and binding agents such as wheat. Water is added and the resulting paste is extruded through holes in a metal plate. The diameter of the holes sets the diameter of the pellets, which can range from less than a millimetre to over a centimetre. As the feed is extruded it is cut to form pellets of the required length. The pellets are dried and oils are added. Adjusting parameters such as temperature and pressure enables the manufacturers to make pellets that suit different fish farming methods, for example feeds that float or sink slowly and feeds suited to recirculation systems. The dry feed pellets are stable for relatively long periods, for convenient storage and distribution. Feeds are delivered in bulk, in large bags—usually one tonne, or in 25 kilogram bags. Smaller quantities of specialist feeds are supplied for use in fish hatcheries.

Fish Disease and Parasites

Like humans and other animals, fish suffer from diseases and parasites. Fish defences against disease are specific and non-specific. Non-specific defences include skin and scales, as well as the mucus layer secreted by the epidermis that traps microorganisms and inhibits their growth. If pathogens breach these defences, fish can develop inflammatory responses that increase the flow of blood to infected areas and deliver white blood cells that attempt to destroy the pathogens.

Specific defences are specialised responses to particular pathogens recognised by the fish's body, that is adaptative immune responses. In recent years, vaccines have become widely used in aquaculture and ornamental fish, for example vaccines for furunculosis in farmed salmon and koi herpes virus in koi. Some commercially important fish diseases are VHS, ich and whirling disease.

Disease

All fish carry pathogens and parasites. Usually this is at some cost to the fish. If the cost is sufficiently high, then the impacts can be characterised as a disease. However disease in fish is not understood well. What is known about fish disease often relates to aquaria fish, and more recently, to farmed fish.

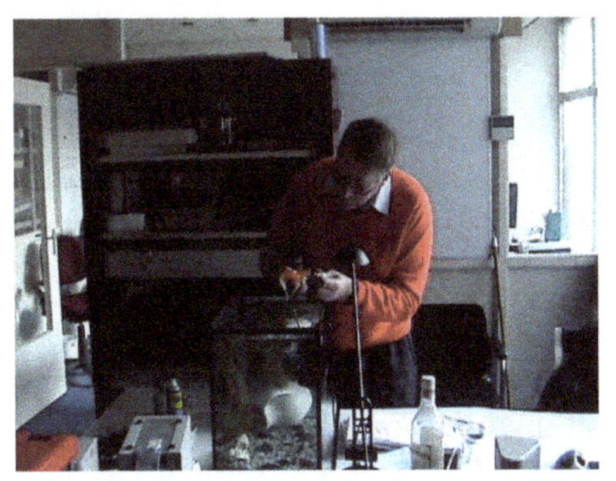

A veterinarian gives an injection to a goldfish

Disease is a prime agent affecting fish mortality, especially when fish are young. Fish can limit the impacts of pathogens and parasites with behavioural or biochemical means, and such fish have reproductive advantages. Interacting factors result in low grade infection becoming fatal diseases. In particular, things that causes stress, such as natural droughts or pollution or predators, can precipitate outbreak of disease.

Disease can also be particularly problematic when pathogens and parasites carried by introduced species affect native species. An introduced species may find invading easier if potential predators and competitors have been decimated by disease.

Pathogens which can cause fish diseases comprise:

- Viral infections, such as esocid lymphosarcoma found in *Esox* species.
- Bacterial infections, such as *Pseudomonas fluorescens* leading to fin rot and fish dropsy.
- Fungal infections.
- Water mould infections, such as *Saprolegnia* sp.
- Metazoan parasites, such as copepods.
- Unicellular parasites, such as *Ichthyophthirius multifiliis* leading to ich.
- Certain parasites like Helminths for example *Eustrongylides*.

Parasites

The isopod Anilocra gigantea parasitising the snapper Pristipomoides filamentosus

Cymothoa exigua is a parasitic crustacean which enters a fish through its gills and destroys the fish's tongue

Parasites in fish are a common natural occurrence. Parasites can provide information about host population ecology. In fisheries biology, for example, parasite communities can be used to distinguish distinct populations of the same fish species co-inhabiting a region. Additionally, parasites possess a variety of specialized traits and life-history strategies that enable them to colonize hosts. Understanding these aspects of parasite ecology, of interest in their own right, can illuminate parasite-avoidance strategies employed by hosts.

Usually parasites (and pathogens) need to avoid killing their hosts, since extinct hosts can mean extinct parasites. Evolutionary constraints may operate so parasites avoid killing their hosts, or the natural variability in host defensive strategies may suffice to keep host populations viable. Parasite infections can impair the courtship dance of male threespine sticklebacks. When that happens, the females reject them, suggesting a strong mechanism for the selection of parasite resistance."

However, not all parasites want to keep their hosts alive, and there are parasites with multistage life cycles who go to some trouble to kill their host. For example, some tapeworms make some fish behave in such a way that a predatory bird can catch it. The predatory bird is the next host for the parasite in the next stage of its life cycle. Specifically, the tapeworm *Schistocephalus solidus* turns infected threespine stickleback white, and then makes them more buoyant so that they splash along at the surface of the water, becoming easy to see and easy to catch for a passing bird.

Parasites can be internal (endoparasites) or external (ectoparasites). Some internal fish parasites are spectacular, such as the philometrid nematode *Philometra fasciati* which is parasitic in the ovary of female Blacktip grouper; the adult female parasite is a red worm which can reach up to 40 centimetres in length, for a diameter of only 1.6 millimetre; the males are tiny. Other internal parasites are found living inside fish gills, include encysted adult didymozoid trematodes, a few trichosomoidid nematodes of the genus *Huffmanela*, including *Huffmanela ossicola* which lives within the gill bone, and the encysted parasitic turbellarian *Paravortex*. Various protists and Myxosporea are also parasitic on gills, where they form cysts.

Fish gills are also the preferred habitat of many external parasites, attached to the gill but living out of it. The most common are monogeneans and certain groups of parasitic copepods, which can be extremely numerous. Other external parasites found on gills are leeches and, in seawater, larvae of gnathiid isopods. Isopod fish parasites are mostly external and feed on blood. The larvae of the Gnathiidae family and adult cymothoidids have piercing and sucking mouthparts and clawed limbs adapted for clinging onto their hosts. *Cymothoa exigua* is a parasite of various marine fish. It causes the tongue of the fish to atrophy and takes its place in what is believed to be the first instance discovered of a parasite functionally replacing a host structure in animals.

Other parasitic disorders, include Gyrodactylus salaris, Ichthyophthirius multifiliis, cryptocaryon, velvet disease, Brooklynella hostilis, Hole in the head, Glugea, Ceratomyxa shasta, Kudoa thyrsites, Tetracapsuloides bryosalmonae, Cymothoa exigua, leeches, nematode, flukes, carp lice and salmon lice.

Although parasites are generally considered to be harmful, the eradication of all parasites would not necessarily be beneficial. Parasites account for as much as or more than half of life's diversity; they perform an important ecological role (by weakening prey) that ecosystems would take some time to adapt to; and without parasites organisms may eventually tend to asexual reproduction, diminishing the diversity of sexually dimorphic traits. Parasites provide an opportunity for the transfer of genetic material between species. On rare, but significant, occasions this may facilitate evolutionary changes that would not otherwise occur, or that would otherwise take even longer.

Below are some life cycles of fish parasites:

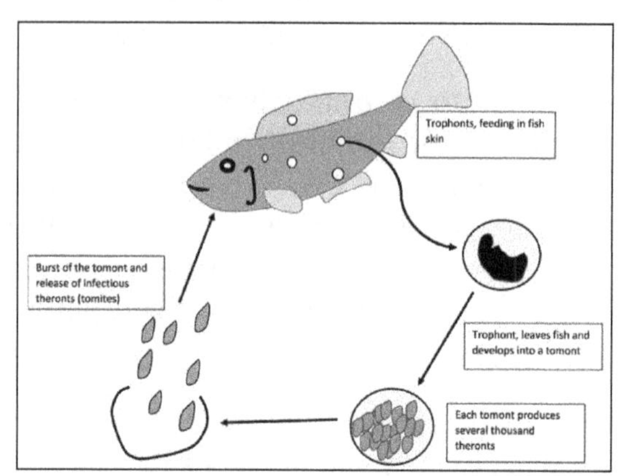

Life cycle of the endoparasite Ichthyophthirius multifiliis, commonly called ich

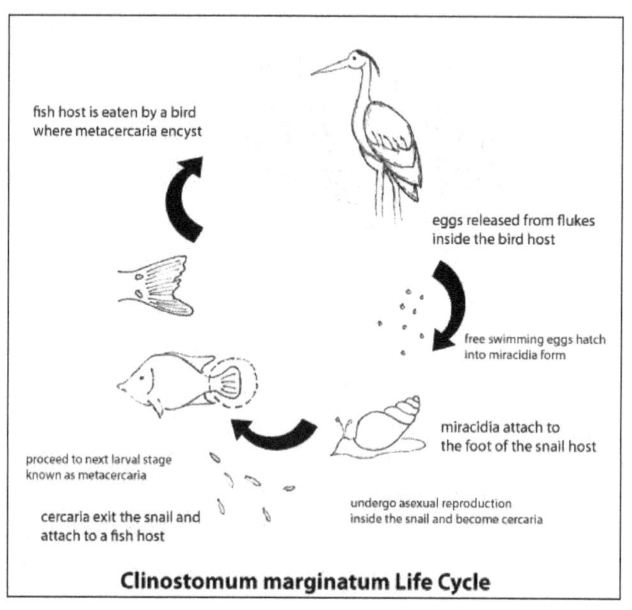

Life cycle of the parasitic fluke *Clinostomum marginatum*, commonly called the yellow grub

Cleaner Fish

Some fish take advantage of cleaner fish for the removal of external parasites. The best known of these are the Bluestreak cleaner wrasses of the genus *Labroides* found on coral reefs in the Indian Ocean and Pacific Ocean. These small fish maintain so-called "cleaning stations" where other fish, known as hosts, will congregate and perform specific movements to attract the attention of the cleaner fish. Cleaning behaviours have been observed in a number of other fish groups, including an interesting case between two cichlids of the same genus, *Etroplus maculatus*, the cleaner fish, and the much larger *Etroplus suratensis*, the host.

More than 40 species of parasites may reside on the skin and internally of the ocean sunfish, motivating the fish to seek relief in a number of ways. In temperate regions, drifting kelp fields harbour cleaner wrasses and other fish which remove parasites from the skin of visiting sunfish. In the tropics, the *mola* will solicit cleaner help from reef fishes. By basking on its side at the surface, the sunfish also allows seabirds to feed on parasites from their skin. Sunfish have been reported to breach more than ten feet above the surface, possibly as another effort to dislodge parasites on the body.

Two cleaner wrasses, Labroides phthirophagus,
servicing a goatfish, Mulloidichthys flavolineatus

Mass Die Offs

Some diseases result in mass die offs. One of the more bizarre and recently discovered diseases produces huge fish kills in shallow marine waters. It is caused by the ambush predator dinoflagellate *Pfiesteria piscicida*. When large numbers of fish, like shoaling forage fish, are in confined situations such as shallow bays, the excretions from the fish encourage this dinoflagellate, which is not normally toxic, to produce free-swimming zoospores. If the fish remain in the area, continuing to provide nourishment, then the zoospores start secreting a neurotoxin. This toxin results in the fish developing bleeding lesions, and their skin flakes off in the water. The dinoflagellates then eat the blood and flakes of tissue while the affected fish die. Fish kills by this dinoflagellate are common, and they may also have been responsible for kills in the past which were thought to have had other causes. Kills like these can be viewed as natural mechanisms for regulating the population of exceptionally abundant fish. The rate at which the kills occur increases as organically polluted land runoff increases.

Wild Salmon

According to Canadian biologist Dorothy Kieser, protozoan parasite *Henneguya salminicola* is commonly found in the flesh of salmonids. It has been recorded in the field samples of salmon returning to the Queen Charlotte Islands. The fish responds by walling off the parasitic infection into a number of cysts that contain milky fluid. This fluid is an accumulation of a large number of parasites.

Henneguya salminicola, a parasite commonly found in the flesh of salmonids on the West Coast of Canada. Coho salmon

Henneguya and other parasites in the myxosporean group have a complex lifecycle where the salmon is one of two hosts. The fish releases the spores after spawning. In the *Henneguya* case, the spores enter a second host, most likely an invertebrate, in the spawning stream. When juvenile salmon out-migrate to the Pacific Ocean, the second host releases a stage infective to salmon. The parasite is then carried in the salmon until the next spawning cycle. The myxosporean parasite that causes whirling disease in trout, has a similar lifecycle. However, as opposed to whirling disease, the *Henneguya* infestation does not appear to cause disease in the host salmon – even heavily infected fish tend to return to spawn successfully.

According to Dr. Kieser, a lot of work on *Henneguya salminicola* was done by scientists at the Pacific Biological Station in Nanaimo in the mid-1980s, in particular, an overview report which states that "the fish that have the longest fresh water residence time as juveniles have the most noticeable infections. Hence in order of prevalence coho are most infected followed by sockeye, chinook, chum and pink." As well, the report says that, at the time the studies were conducted, stocks from the middle and upper reaches of large river systems in British Columbia such as Fraser, Skeena, Nass and from mainland coastal streams in the southern half of B.C. "are more likely to have a low prevalence of infection." The report also states "It should be stressed that *Henneguya*, economically deleterious though it is, is harmless from the view of public health. It is strictly a fish parasite that cannot live in or affect warm blooded animals, including man".

According to Klaus Schallie, Molluscan Shellfish Program Specialist with the Canadian Food Inspection Agency, "*Henneguya salminicola* is found in southern B.C. also and in all species of salmon. I have previously examined smoked chum salmon sides that were riddled with cysts and some sockeye runs in Barkley Sound (southern B.C., west coast of Vancouver Island) are noted for their high incidence of infestation."

Sample of pink salmon infected with Henneguya salminicola, caught off Haida Gwaii, Western Canada in 2009

Sea lice, particularly *Lepeophtheirus salmonis* and a variety of *Caligus* species, including *Caligus clemensi* and *Caligus rogercresseyi*, can cause deadly infestations of both farm-grown and wild salmon. Sea lice are ectoparasites which feed on mucous, blood, and skin, and migrate and latch onto the skin of wild salmon during free-swimming, planktonic *naupli* and *copepodid* larval stages, which can persist for several days. Large numbers of highly populated, open-net salmon farms can create exceptionally large concentrations of sea lice; when exposed in river estuaries containing large numbers of open-net farms, many young wild salmon are infected, and do not survive as a result. Adult salmon may survive otherwise critical numbers of sea lice, but small, thin-skinned juvenile salmon migrating to sea are highly vulnerable. On the Pacific coast of Canada, the louse-induced mortality of pink salmon in some regions is commonly over 80%.

Farmed Salmon

Atlantic salmon

Gyrodactylus salaris, also called salmon fluke, a monogenean parasite, spread from Norwegian hatcheries to wild salmon, and devastated some wild salmon populations.

infectious salmon anemia (ISAv) was discovered in Norway in an Atlantic salmon hatchery. Eighty percent of the fish in the outbreak died. ISAv, a viral disease, is now a major threat to the viability of Atlantic salmon farming. It is now the first of the diseases classified on List One of the European Commission's fish health regime. Amongst other measures, this requires the total eradication of the entire fish stock should an outbreak of the disease be confirmed on any farm. ISAv seriously affects salmon farms in Chile,

Norway, Scotland and Canada, causing major economic losses to infected farms. As the name implies, it causes severe anemia of infected fish. Unlike mammals, the red blood cells of fish have DNA, and can become infected with viruses. The fish develop pale gills, and may swim close to the water surface, gulping for air. However, the disease can also develop without the fish showing any external signs of illness, the fish maintain a normal appetite, and then they suddenly die. The disease can progress slowly throughout an infected farm and, in the worst cases, death rates may approach 100 percent. It is also a threat to the dwindling stocks of wild salmon. Management strategies include developing a vaccine and improving genetic resistance to the disease.

In the wild, diseases and parasites are normally at low levels, and kept in check by natural predation on weakened individuals. In crowded net pens they can become epidemics. Diseases and parasites also transfer from farmed to wild salmon populations. A recent study in British Columbia links the spread of parasitic sea lice from river salmon farms to wild pink salmon in the same river." The European Commission concluded "The reduction of wild salmonid abundance is also linked to other factors but there is more and more scientific evidence establishing a direct link between the number of lice-infested wild fish and the presence of cages in the same estuary." It is reported that wild salmon on the west coast of Canada are being driven to extinction by sea lice from nearby salmon farms. Antibiotics and pesticides are often used to control the diseases and parasites.

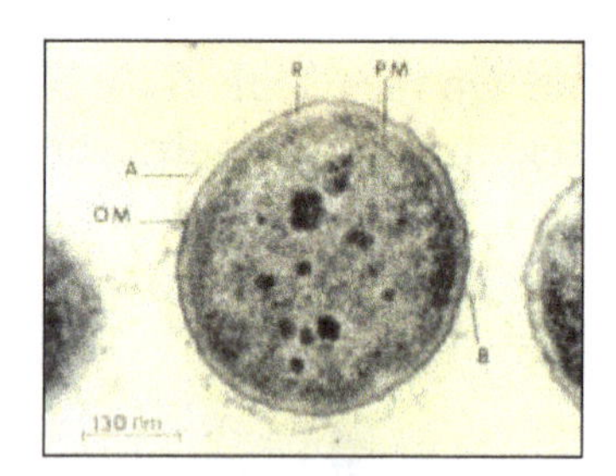

Aeromonas salmonicida, a Gram-negative bacteria, causes the disease furunculosis in marine and freshwater fish

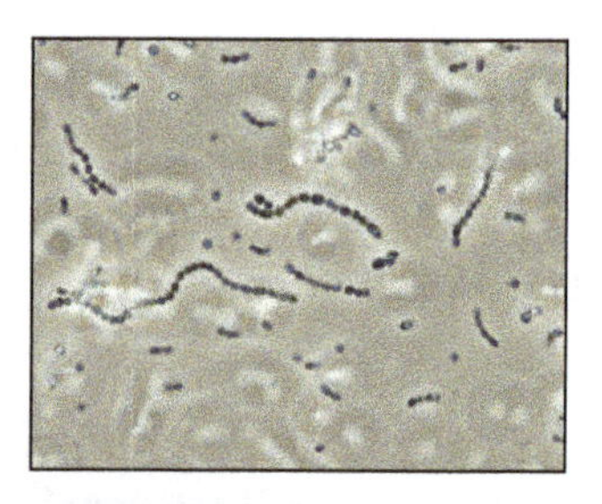

Streptococcus iniae, a Gram-positive, sphere-shaped bacteria caused losses in farmed marine and freshwater finfish of US$100 million

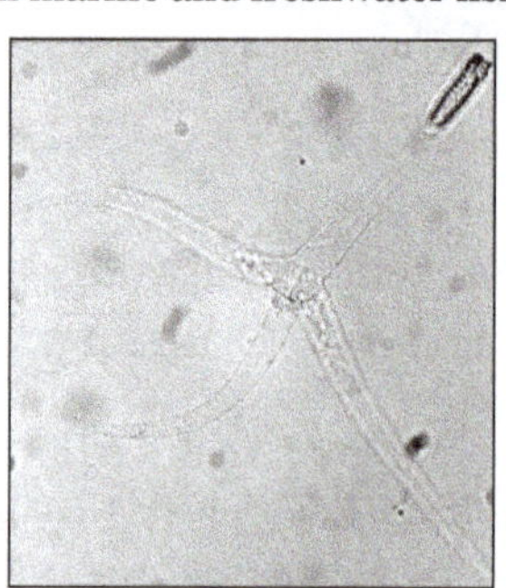

Myxobolus cerebralis, a myxosporean parasite, causes whirling disease in farmed salmon and trout and also in wild fish populations

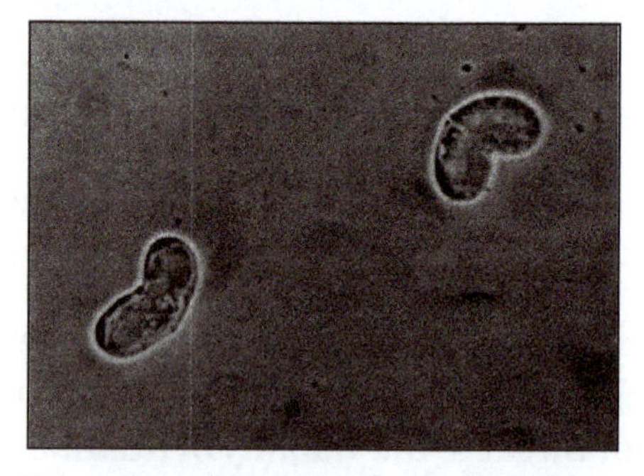

Ceratomyxa shasta, another myxosporean parasite, infects salmonid fish on the Pacific coast of North America

Coral Reef Fish

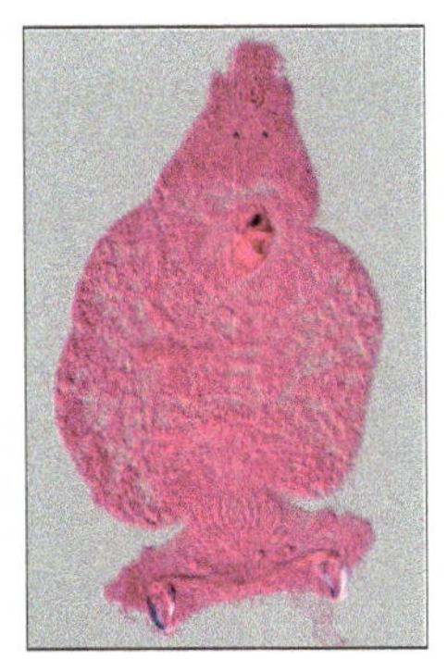

Monogenean parasite on the gill of a grouper

Coral reef fish are characterized by high biodiversity. As a consequence parasites of coral reef fish show tremendous variety. Parasites of coral reef fish include nematodes, Platyhelminthes (cestodes, digeneans, and monogeneans), leeches, parasitic crustaceans such as isopods and copepods, and various microorganisms such as myxosporidia and microsporidia. Some of these fish parasites have heteroxenous life cycles (i.e. they have several hosts) among which sharks (certain cestodes) or molluscs (digeneans). The high biodiversity of coral reefs increases the complexity of the interactions between parasites and their various and numerous hosts. Numerical estimates of parasite biodiversity have shown that certain coral fish species have up to 30 species of parasites. The mean number of parasites per fish species is about ten. This has a consequence in term of co-extinction. Results obtained for the coral reef fish of New Caledonia suggest that extinction of a coral reef fish species of average size would eventually result in the co-extinction of at least ten species of parasites.

Aquarium Fish

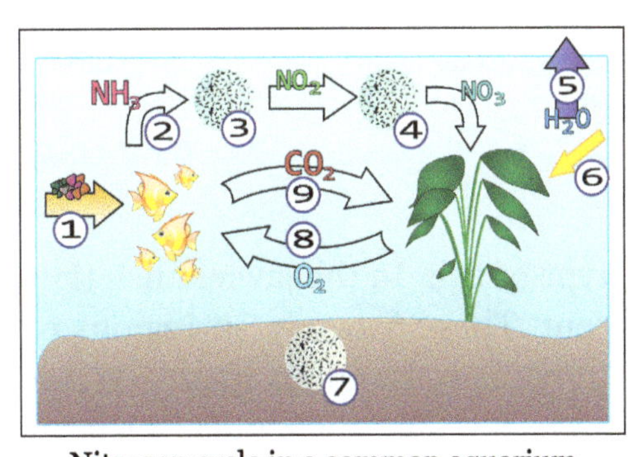

Nitrogen cycle in a common aquarium

Ornamental fish kept in aquariums are susceptible to numerous diseases. In most aquarium tanks, the fish are at high concentrations and the volume of water is limited. This means that communicable diseases can spread rapidly to most or all fish in a tank. An improper nitrogen cycle, inappropriate aquarium plants and potentially harmful freshwater invertebrates can directly harm or add to the stresses on ornamental fish in a tank. Despite this, many diseases in captive fish can be avoided or prevented through proper water conditions and a well-adjusted ecosystem within the tank. Ammonia poisoning is a common disease in new aquariums, especially when immediately stocked to full capacity.

Goldfish with dropsy

Columnaris in the gill of a chinook salmon

Due to their generally small size and the low cost of replacing diseased or dead aquarium fish, the cost of testing and treating diseases is often seen as more trouble than the value of the fish.

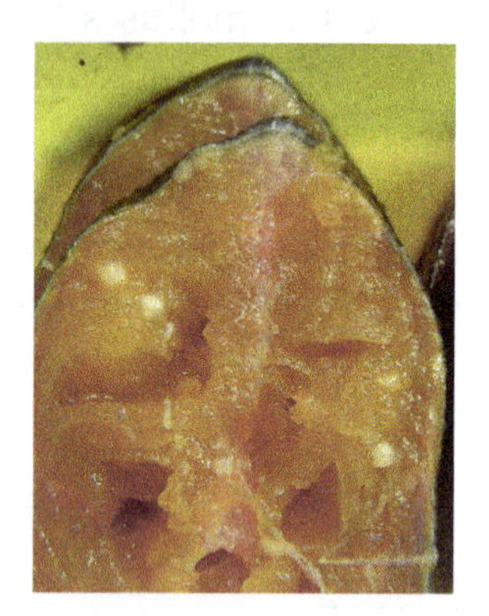

The parasite *Henneguya zschokkei* in salmon beard

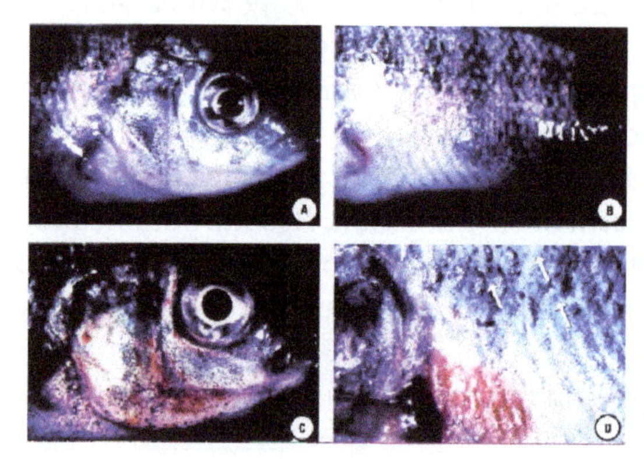

Skin ulcers in tilapia exposed to Pfiesteria shumwayae

Immune System

Immune organs vary by type of fish. In the jawless fish (lampreys and hagfish), true lymphoid organs are absent. These fish rely on regions of lymphoid tissue within other organs to produce immune cells. For example, erythrocytes, macrophages and plasma cells are produced in the anterior kidney (or pronephros) and some areas of the gut (where granulocytes mature.) They resemble primitive bone marrow in hagfish. Cartilaginous fish (sharks and rays) have a more advanced immune system. They have three specialized organs that are unique to chondrichthyes; the epigonal organs (lymphoid tissue similar to mammalian bone) that surround the gonads, the Leydig's organ within the walls of their esophagus, and a spiral valve in their intestine. These organs house typical immune cells (granulocytes, lymphocytes and plasma cells). They also possess an identifiable thymus and a well-developed spleen (their most important immune organ) where various lymphocytes, plasma cells and macrophages develop and are stored. Chondrostean fish (sturgeons, paddlefish and bichirs) possess a major site for the production of granulocytes within a mass that is associated with the meninges (membranes surrounding the central nervous system.) Their heart is frequently covered with tissue that contains lymphocytes, reticular cells and a small number of macrophages. The chondrostean kidney is an important hemopoietic organ; where erythrocytes, granulocytes, lymphocytes and macrophages develop.

Like chondrostean fish, the major immune tissues of bony fish (or teleostei) include the kidney (especially the anterior kidney), which houses many different immune cells. In addition, teleost fish possess a thymus, spleen and scattered immune areas within mucosal tissues (e.g. in the skin, gills, gut and gonads). Much like the mammalian immune system, teleost erythrocytes, neutrophils and granulocytes are believed to reside in the spleen whereas lymphocytes are the major cell type found in the thymus. In 2006, a lymphatic system similar to that in mammals was described in one species of teleost fish, the zebrafish. Although not confirmed as yet, this system presumably will be where naive (unstimulated) T cells accumulate while waiting to encounter an antigen.

Spreading Disease and Parasites

The capture, transportation and culture of bait fish can spread damaging organisms between ecosystems, endangering them. In 2007, several American states, including Michigan, enacted regulations designed to slow the spread of fish diseases, including viral hemorrhagic septicemia, by bait fish. Because of the risk of transmitting *Myxobolus cerebralis* (whirling disease), trout and salmon should not be used as bait. Anglers may increase the possibility of contamination by emptying bait buckets into fishing venues and collecting or using bait improperly. The transportation of fish from one location to another can break the law and cause the introduction of fish and parasites alien to the ecosystem.

Eating Raw Fish

Differential symptoms of parasite infection by raw fish: Clonorchis sinensis (a trematode/fluke), Anisakis (a nematode/roundworm) and Diphyllobothrium a (cestode/tapeworm), all have gastrointestinal, but otherwise distinct, symptoms.

Though not a health concern in thoroughly cooked fish, parasites are a concern when human consumers eat raw or lightly preserved fish such as sashimi, sushi, ceviche, and gravlax. The popularity of such raw fish dishes makes it important for consumers to be aware of this risk. Raw fish should be frozen to an internal temperature of −20 °C

(−4 °F) for at least 7 days to kill parasites. It is important to be aware that home freezers may not be cold enough to kill parasites.

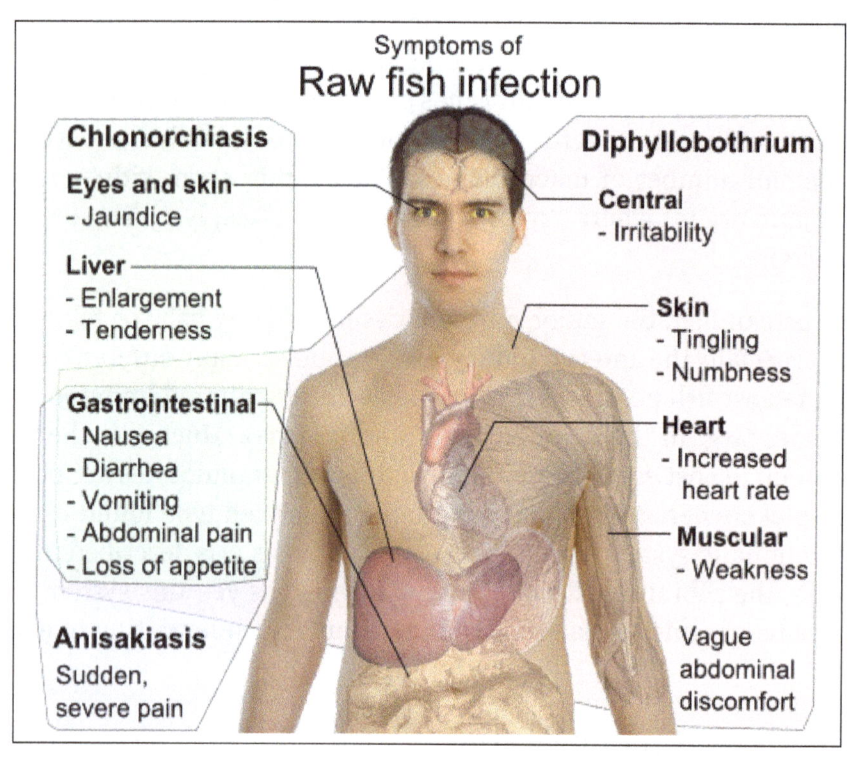

Traditionally, fish that live all or part of their lives in fresh water were considered unsuitable for sashimi due to the possibility of parasites. Parasitic infections from freshwater fish are a serious problem in some parts of the world, particularly Southeast Asia. Fish that spend part of their life cycle in salt water, like salmon, can also be a problem. A study in Seattle, Washington showed that 100% of wild salmon had roundworm larvae capable of infecting people. In the same study farm raised salmon did not have any roundworm larvae.

Parasite infection by raw fish is rare in the developed world (fewer than 40 cases per year in the U.S.), and involves mainly three kinds of parasites: Clonorchis sinensis (a trematode/fluke), Anisakis (a nematode/roundworm) and Diphyllobothrium (a cestode/tapeworm). Infection by the fish tapeworm *Diphyllobothrium latum* is seen in countries where people eat raw or undercooked fish, such as some countries in Asia, Eastern Europe, Scandinavia, Africa, and North and South America. Infection risk of anisakis is particularly higher in fishes which may live in a river such as salmon (*shake*) in Salmonidae, mackerel (*saba*). Such parasite infections can generally be avoided by boiling, burning, preserving in salt or vinegar, or freezing overnight. Even Japanese people never eat raw salmon or ikura (salmon roe), and even if they seem raw, these foods are not raw but are frozen overnight to prevent infections from parasites, particularly anisakis.

Below are some life cycles of fish parasites that can infect humans:

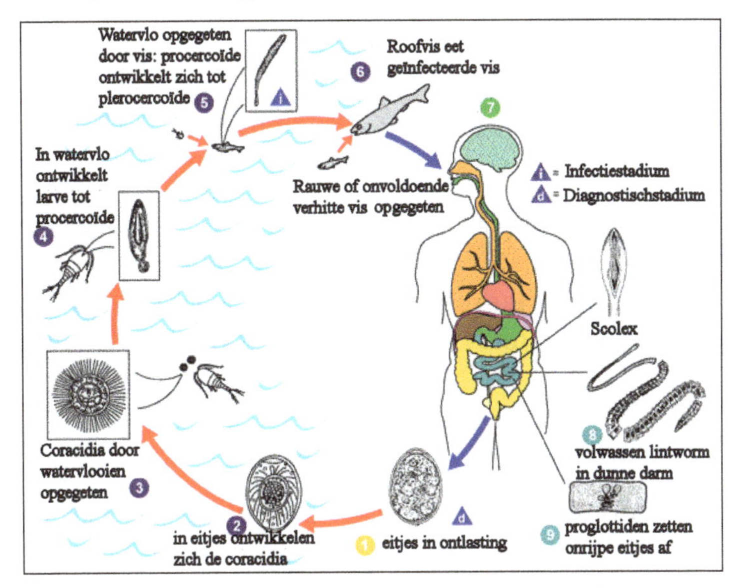

Life cycle of the fish tapeworm Diphyllobothrium latum

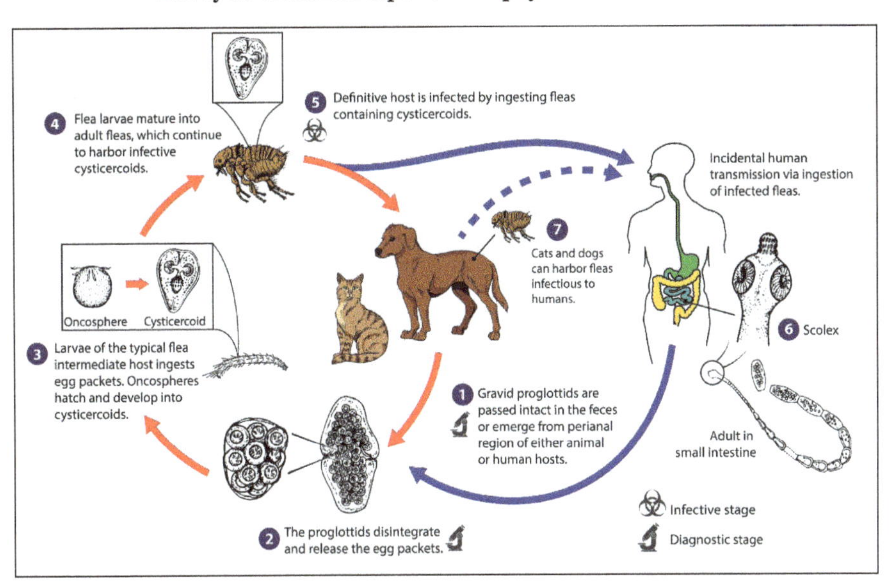

Life cycle of the digenean Metagonimus, an intestinal fluke

Fish Farming Issues

Factory fish farming — also known as aquaculture — is generally big, dirty, and dangerous, just like factory farming on land. Around half of the seafood eaten in the entire world comes from these types of facilities as producers attempt to produce fish as cheaply as possible. Massive amounts of antibiotics, hormones, and pesticides are required to keep disease at bay just to keep fish and shrimp alive in overcrowded conditions (typically in nets, cages, or ponds). The risk of contamination is high, both to the surrounding water and within the enclosures themselves. Multinational corporations have forever changed the way food is grown on land to the detriment of public health, the environment, local communities and food quality itself, and they are poised to do the same in the water.

Unfortunately, even though people have become increasingly conscious about the environmental, cultural and economic repercussions of their seafood choices, the U.S. government continues to push for the development of open ocean aquaculture. The federal government has already spent millions to promote this troubled industry, despite poor results.

Bad for the Environment

Uneaten fish feed, fish waste, and any antibiotics or chemicals used in fish farm operations flow through the cages directly into the ocean. This can significantly harm the ocean environment. Caged fish can escape and compete for resources or interbreed with wild fish and weaken important genetic traits. Farmed fish can spread disease to wild fish.

Further, factory fish farms tend to grow top-of-the-food-chain carnivorous fish that eat small, wild fish — it can take several pounds of wild fish to grow one pound of farmed fish. This undermines the wild marine food chain.

Bad for Communities

Factory fish farms may interfere with the livelihoods of commercial and recreational fishermen by displacing them from traditional fishing grounds or harming wild fish populations. Flooding the market with cheap farmed fish can drive down prices for wild fish, putting fishermen out of business and fishing communities in peril.

Bad for our Health

Fish produced at factory fish farms can have higher levels of contaminants than wild fish, which may lead to health risks for consumers. And the use of antibiotics on fish farms can cause drug-resistant bacteria to develop, which may then be passed on to humans.

Reasons to Stop Ocean Fish Farming

- Ocean fish farming facilities can lead to conflict in areas including fishing grounds and routes to those fishing grounds, vessel traffic lanes, military sites, marine reserves and sanctuaries, protected and fragile areas and areas of significant multiple use.

- Economic concerns include loss of jobs for commercial fishermen, plummeting fish prices as cheap fish flood the market and decreased opportunities for recreational fishing as water and fish become polluted from nearby factory fish farms.
- Escapes are a given due to complications like severe weather, sharks and other predators, equipment failure, and human error. Fish escapes can jeopardize the recovery of depleted or endangered species and lead to the spread of diseases, breeding with wild populations and the disruption of natural ecosystems.
- Habitats can be severely impacted by dredging, drilling, dropping large anchors, the introduction of new predators and sediment disturbances.
- Water pollution from offshore fish farms can include fish waste, excess food, fish escapes, antibiotics and various chemicals from fish farms, resulting in water pollution and poisoning of surrounding habitats.

References

- Fish-farming, energy-government-and-defense-magazines, environment: encyclopedia.com, Retrieved 29 June, 2019
- "Stocking". Pennsylvania Fish and Boat Commission. Retrieved 10 March 2019
- Trout Stocking Summary". Pennsylvania Fish & Boat Commission. Retrieved 11 March 2019
- Edmonds, Molly. "How Fish Stocking Works". How Stuff Works. Retrieved 10 March 2019
- "Recreational Fishing - Statistics & Facts". Statista. Retrieved 11 March 2019
- Faulk, Cynthia K.; Holt, G. Joan (2003). "Lipid Nutrition and Feeding of Cobia Rachycentron canadum Larvae". Journal of the World Aquaculture Society. 34 (3): 368–378. Doi:10.1111/j.1749-7345.2003.tb00074.x
- "Fish Farming Information and Resources". Farms.com. Retrieved November 25, 2018
- "Cultured Aquatic Species Information Programme Salmo trutta". Food and Agriculture Organization of the United Nations. Retrieved 2015-01-21
- Nash, Colin E.; Waknitz, F.William (2003). "Interactions of Atlantic salmon in the Pacific Northwest". Fisheries Research. 62 (3): 237–254. Doi:10.1016/S0165-7836(03)00063-8. ISSN 0165-7836
- Fairfield, T (2000). A commonsense guide to fish health. Woodbury, N.Y: Barron's Educational Series. ISBN 978-0-7641-1338-3
- Moyle, PB and Cech, JJ (2004) Fishes, An Introduction to Ichthyology. 5th Ed, Benjamin Cummings. ISBN 978-0-13-100847-2
- Factory-fish-farming, insight: foodandwaterwatch.org, Retrieved 23 July, 2019

Algaculture 6

- **Aquaculture of Giant Kelp**
- **Kelp Forest**
- **Culture of Microalgae in Hatcheries**
- **Photobioreactor**
- **Seaweed Farming**
- **Algae Bioreactor**
- **Algae Harvesting**

Algaculture is a branch of aquaculture that deals with the farming of algae species. They are broadly divided into two main categories, namely, microalgae and macroalgae. The diverse aspects of algaculture as well as the diverse uses for algae have been thoroughly discussed in this chapter.

Algaculture is the commercial cultivation of algae. Algae are simple green plants that grow in water. Their green color means they produce their own food using photosynthesis, just like grass, trees and corn. Algae come in two main forms. Macroalgae are seaweeds. Kelp grows to more than 180 feet (55 meters) long in the ocean. Nori is the variety you'll find wrapped around your sushi. Microalgae are tiny, single-celled plants that float in the water, each one visible only through a microscope.

Algaculture is nothing new. Seaweed was first cultivated in Japan at least 1,500 years ago and algae production is still a big business there. Dulse has long been eaten in the British Isles and the microalgae spirulina were harvested by the Aztecs of 16th-century Mexico. In addition to providing human food, seaweeds have been used for fertilizers. They provide the food thickener carrageen and other gelling agents and stabilizers that show up in everything from soup to toothpaste. Worldwide, algae production is a $6 billion business.

Today, algae are attracting new interest and resarch investment because of their potential to provide energy and combat environmental threats. Part of the organic mass of algae takes the form of oil, which can be squeezed out and converted to biodiesel fuel.

Algae beat land plants hands down in the efficiency with which they produce oil. Some varieties of algae yield an oil that can be refined into gasoline and even jet fuel. The carbohydrate portion of the plants can be fermented for ethanol production.

Algae can convert waste carbon dioxide, a greenhouse gas that pours from smokestacks, to usable products. They can help clean dirty water, converting pollutants to biomass. They have additional uses in pharmaceuticals and cosmetics.

With all this potential, this "weed" certainly seems to deserve a closer look.

The Promise of Algae

Why have algae generated excitement and attracted research investment in recent years? Like all green plants, algae contain chloroplasts in their cells. These tiny structures are charged with chlorophyll, a molecule that uses light energy to combine carbon and water into a simple sugar. The cells further transform some of these sugars into proteins and lipids or oil.

But if algae are doing the same thing as corn, wheat and apple trees, why bother raising them? After all, corn on the cob, sweet rolls and apple pie taste better than seaweed to most of us. Here are some of the things algae have going for them:

- Productivity: Algae are super fast-growing. Land plants take months or years to reach maturity. Algae can complete their entire life cycle in a single day. Some algae can double their biomass in just an hour.
- Efficiency: When it comes to converting solar energy to biomass, algae are all business. Because they're supported by and take their nutrients directly from water, they need no roots, stems or flowers. Land plants use as much as 95 percent of their energy building the structures they need for support, feeding and breeding.
- Concentration: Because of their efficiency, algae can be grown in a very concentrated space. They produce up to 100 times more oil per acre than land plants.
- Versatility: It's estimated that there are more than 70,000 species of algae, many of them not yet classified. Growers can pick ones suited to conditions and goals, selecting varieties for a specific temperature range or water salinity, for example.
- Non-competition: Algae don't compete with current crops for land or fresh water. They can be grown in ponds in locations, like deserts, that don't sustain land plants. Some varieties prefer saline or polluted water.

Attracted by all these advantages, algae cultivators have been working diligently to come up with efficient and economical ways to grow and harvest the plants. The cost factor is currently the great challenge that must be overcome to make algae commercially viable.

Commercial Cultivation of Algae

All algaculture requires three basics: water, light and nutrients.

Water It doesn't need to be potable; different types of algae grow nicely in fresh water, salt water and dirty water. Sunlight, because it's free, is the preferred light. But sunlight reaches only 3 or 4 inches (7 to 10 centimeters) into a mass of algae, so growers must agitate the algae to expose all of it to the light. The main nutrient is carbon dioxide, which can come from the air or other source. Agitation or bubbling dissolves it into the water. The grower must supply other nutrients, like nitrogen and trace elements, if they aren't already in the water.

There are three basic systems for cultivating algae, each with its advantages and disadvantages:

- Open pond: The simplest and cheapest way to grow algae is in large, shallow ponds. The water is often divided into concentric lanes or raceways, with paddlewheels to move the algae mixture in a circle. This helps bring algae to the surface, where they're exposed to light, and mixes nutrients and carbon dioxide into the liquid. The open-pond method produces less algae biomass than other methods. It loses water to evaporation, so more must be added. And it allows for contamination by predators or undesirable algae.
- Closed pond: This method is similar to an open pond, but the water is covered by a Plexiglas greenhouse. This raises the cost, but allows greater control of the process. It reduces evaporation and contamination and extends the growing season. Growers can raise algae year-round if the space is heated.
- Biophotoreactor: A completely closed system, the biophotoreactor consists of glass or acrylic tubes where the algae are exposed to light. Pumps move the water, nutrients and algae through the tubes and storage tanks. Some reactors automatically harvest the algae when they're ready. This approach gives growers the most control over the process and the most efficient way to produce algae biomass. But it's also the most costly to set up and operate.

All of these systems are designed for growing microalgae, the one-celled varieties that float in water. Growers usually cultivate macroalgae in the open sea. The water already contains the nutrients the algae need and provides good growing conditions. The traditional method was simply to harvest wild seaweed, and this is still done in coastal areas around the world.

With increased demand, growers began to cultivate seaweed. For some varieties, such as kelp, spores are attached to ropes that are then anchored in the ocean and the seaweed is allowed to grow. Other types grow from pieces of seaweed that are fixed to nets or deposited in pools.

Agriculture has been around for 10,000 years. Algaculture is relatively new. Scientists and engineers are actively studying the best ways to raise algae efficiently. The harvesting of plants is another subject of intense research.

Uses for Algae

it's an ideal source of renewable energy and could be the ultimate green fuel.

Research by the U.S. government and companies like Boeing, Chevron and Honeywell are developing ways to make algaculture an economically viable foundation for a new generation of energy. Part of the attraction is the range of fuels into which algae can be converted.

- Biodiesel is the simplest way to tap algae's energy potential. Like any vegetable oil, oil from algae can be chemically transformed into biodiesel fuel. Compared to land plants like soybeans or corn, algae use less land and fresh water, grow faster and have higher concentrations of oil.
- Refined transportation fuels are another area of promise for algae. Some algae produce oil that can be refined into gasoline or even jet fuel, and without the sulfur and nitrogen compounds in petroleum. Manufacturers can process it in the same refineries as petroleum-based stock. In 2011, the first commercial jet flight powered by algal oil flew from Houston to Chicago.
- Ethanol, which is commonly added to gasoline, can be produced from algae as well as land plants. Besides oil, algae are made up from carbohydrates and cellulose walls. These materials can be fermented by yeast into ethanol or grain alcohol.
- Methane, the main ingredient in natural gas, is produced when bacteria digest algae. A clean and versatile fuel, methane can be used to produce electricity or power vehicles. It represents another biofuel option for algae.

Algae actually thrive on polluted water, which means they can be used for waste water treatment. Algae turn pollutants from municipal, industrial or agricultural waste water into usable byproducts like animal feed or biomass for conversion to energy. Algae naturally accumulate heavy metals for removal or recycling.

Because carbon dioxide, the greenhouse gas that contributes to climate change, is algae's favorite food, the plants can be used for carbon capture. They convert the gas to organic carbon compounds at a far faster clip than land plants. One pound (453.6 grams) of algae consumes 2 pounds (907.2 grams) of carbon dioxide. Feed the waste gas of a coal-burning power plant into a mass of algae, and they literally eat it up. Waste gas can be stored for permanent elimination from the atmosphere, or converted to fuel to cut the use of fossil fuels.

Algae continue to play a role as human food and supplements. People eat seaweed in salads and sushi and take supplements made from the microalgae spirulina. Algae provide complete protein, omega-3 fatty acids and vitamins. Carageen is extracted from red seaweed known as Irish moss and used as a thickener.

Algae are also being used as feed for cattle and for marine animals like shrimp and shellfish. The biomass left after algae have been processed can sometimes be applied as organic fertilizer to farm fields. Algae find minor uses in cosmetics and pharmaceuticals as well.

Research into growing, harvesting and processing algae is advancing on many fronts. Given its immense value, there's no doubt that this simple "weed" will play a growing role in the future of our society and economy.

Aquaculture of Giant Kelp

Aquaculture of giant kelp, *Macrocystis pyrifera*, is the cultivation of kelp for uses such as food, dietary supplements or potash. Giant kelp contains compounds such as iodine, potassium, other minerals vitamins and carbohydrates.

At the beginning of the 20th century California kelp beds were harvested for their potash. Commercial interest increased during the 1970s and the 1980s due to the production of alginates, and also for biomass production for animal feed due to the energy crisis. However commercial production for *M. pyrifera* never developed. With the end of the energy crisis and the decline in alginate prices, research into farming *Macrocystis* declined.

The supply of *M. pyrifera* for alginate production relied heavily on restoration and management of natural beds during the early 1990s. Other functions such as substrate stabilization were explored in California, where the "Kelp bed project" transplanted 3-6m adult specimens to increase the stability of the harbor and promote diversity.

Research is investigating its use as feed for other aquaculture species such as shrimp.

China and Chile are the largest producers of aquatic plants, each producing over 300,000 tonnes in 2007. How much of this total can be attributed to *M. pyrifera* is unclear. Both countries culture a variety of species, in Chile 50% of the production involves *Phaeophytes* and the other 50% is *Rhodophytes*. China produces a larger variety of seaweeds including chlorophytes. Experiments in Chile are exploring hybrids of *M. pyrifera* and *M. integrifolia*.

Methods

The most common method of cultivating *M. pyrifera* was developed in China in the 1950s. It is called the long line cultivation system, where the sporelings are produced in a cooled water greenhouse and then planted in the ocean attached to long lines. The depth at which they are grown varies. This species alternates generations in its life cycle, cycling between a large sporophyte and a microscopic gametophyte. The sporophyte is harvested as seaweed. The mature sporophytes form the reproductive organs called sori. They are found on the underside of the leaves and produce the motile zoospores that germinate into the gametophyte. To induce sporalation, plants are dried for up to twelve hours and placed in a seeding container filled with seawater of about 9-10 °C; salinity of 30% and a pH of 7.8-7.9. Photoperiod is controlled during sporolation and growth phases. A synthetic twine of about 2 – 6mm in diameter is placed on the bottom of the same container after sporalation. The released zoospores attach to the twine and begin to germinate into male and female gametophytes. Upon maturity these gametophytes release sperm and egg cells that fuse in the water column and attach themselves to the same substrate as the gametophytes (the twine). These plants are reared into young sporophytes for up to 60 days.

These strings are either wrapped around or are cut up into small pieces and attached to a larger diameter cultivation rope. The cultivation ropes vary, but extend approximately 60m with floating buoys attached. The depths vary. In China, *M. pyrifera* is cultivated on the surface with floating buoys attached every 2-3m and the ends of the rope attached to a wooden peg anchored to the substrate. Individual ropes are usually hung at 50 cm intervals. In Chile *M. pyrifera* is grown at a depth of 2m using buoys to keep the plants at a constant depth. These are then let alone to grow until harvest.

Problems that afflict this method include management of the transition from spore to gametophyte and embryonic sporophyte which are done on a terrestrial facility with careful control of water flow, temperature, nutrients and light. The Japanese use a forced cultivation method where 2 years of growth is achieved within a single growing season by controlling inputs.

In China a project for offshore/deep water cultivation used various farm structures to facilitate growth, including pumping nutrients from deep water into the beds. The greatest benefit for this approach was that the algae were released from size constraints of shallow waters. Issues with operational and farm designs plagued deep water cultivation and ended further exploration.

Harvesting

The duration of cultivation varies by region and farming intensity. This species is usually harvested after two growth seasons (2 years). *M. pyrifera* that is artificially cultivated on ropes is harvested by a pulley system that is attached to boats that pull the individual lines on the vessels for cleaning. Other countries such as the US rely primarily on naturally grown *M. pyrifera,* use boats to harvest the surface canopy several times per year. This is possible due to fast growth while the vegetative and reproductive parts are left undamaged.

Applications

In the UK, legislation defines giant kelp as a nuisance, invasive specimens are mechanically removed.

The demand for *M. pyrifera* centers on fertilizers, bioremediation and feed for abalone and sea urchins.

Carbon Sequestration

Offsetting current carbon emissions would require some 50 trillion trees. An alternative offset would be to cultivate kelp forests. Kelp can grow at 2 feet per day, 30 times faster than terrestrial plants. Planting kelp across 9% of the oceans (4.5 x the area of Australia) could provide the same offset. Additionally, the kelp would support a fish harvest of 2 megatons per year and reduce ocean acidification. Large scale open ocean forestry would require engineered substrate and added nutrients.

Biofuel

As of 2017, the Wrigley Institute of Environmental Studies was testing farming of kelp near Catalina Island for conversion to a biofuel by thermochemical liquefaction.

Food

Small-scale cultivation uses kelp as a replacement for kale.

Kelp Forest

Kelp forests are underwater areas with a high density of kelp, which covers about 25% of the world's coastlines. They are recognized as one of the most productive and dynamic ecosystems on Earth. Smaller areas of anchored kelp are called kelp beds.

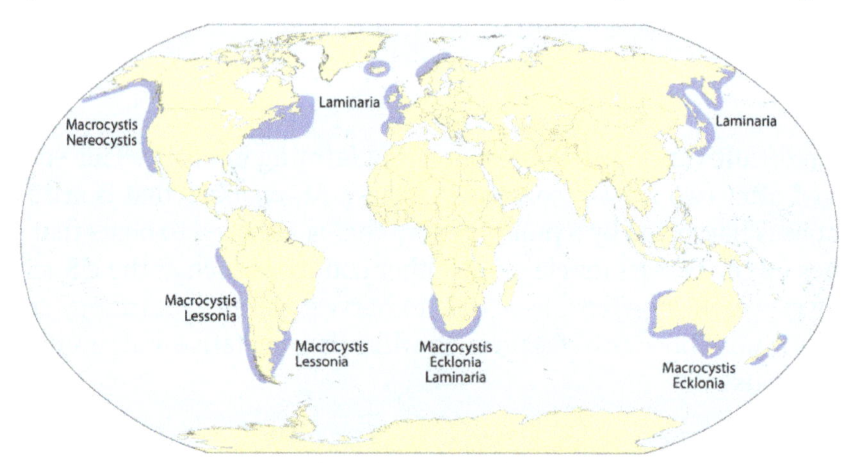

Global distribution of kelp forests

Kelp forests occur worldwide throughout temperate and polar coastal oceans. In 2007, kelp forests were also discovered in tropical waters near Ecuador.

Physically formed by brown macroalgae, kelp forests provide a unique habitat for marine organisms and are a source for understanding many ecological processes. Over the last century, they have been the focus of extensive research, particularly in trophic ecology, and continue to provoke important ideas that are relevant beyond this unique ecosystem. For example, kelp forests can influence coastal oceanographic patterns and provide many ecosystem services.

However, the influence of humans has often contributed to kelp forest degradation. Of particular concern are the effects of overfishing nearshore ecosystems, which can release herbivores from their normal population regulation and result in the overgrazing of kelp and other algae. This can rapidly result in transitions to barren landscapes where relatively few species persist. The implementation of marine protected areas is one management strategy useful for addressing such issues, since it may limit the impacts of fishing and buffer the ecosystem from additive effects of other environmental stressors.

Kelp

The term kelp refers to marine algae belonging to the order Laminariales (phylum: Heterokontophyta). Though not considered a taxonomically diverse order, kelps are highly diverse structurally and functionally. The most widely recognized species are the giant kelps (*Macrocystis* spp.), although numerous other genera such as *Laminaria*, *Ecklonia*, *Lessonia*, *Alaria*, and *Eisenia* are described.

A wide range of sea life uses kelp forests for protection or food, including fish. In the North Pacific kelp forests, particularly rockfish, and many invertebrates, such as amphipods, shrimp, marine snails, bristle worms, and brittle stars. Many marine mammals and birds are also found, including seals, sea lions, whales, sea otters, gulls, terns, snowy egrets, great blue herons, and cormorants, as well as some shore birds.

Frequently considered an ecosystem engineer, kelp provides a physical substrate and habitat for kelp forest communities. In algae (kingdom Protista), the body of an individual organism is known as a thallus rather than as a plant (kingdom Plantae). The morphological structure of a kelp thallus is defined by three basic structural units:

- The holdfast is a root-like mass that anchors the thallus to the sea floor, though unlike true roots it is not responsible for absorbing and delivering nutrients to the rest of the thallus.
- The stipe is analogous to a plant stalk, extending vertically from the holdfast and providing a support framework for other morphological features.
- The fronds are leaf- or blade-like attachments extending from the stipe, sometimes along its full length, and are the sites of nutrient uptake and photosynthetic activity.

In addition, many kelp species have pneumatocysts, or gas-filled bladders, usually located at the base of fronds near the stipe. These structures provide the necessary buoyancy for kelp to maintain an upright position in the water column.

The environmental factors necessary for kelp to survive include hard substrate (usually rock or sand), high nutrients (e.g., nitrogen, phosphorus), and light (minimum annual irradiance dose > 50 E m^{-2}). Especially productive kelp forests tend to be associated with areas of significant oceanographic upwelling, a process that delivers cool, nutrient-rich water from depth to the ocean's mixed surface layer. Water flow and turbulence facilitate nutrient assimilation across kelp fronds throughout the water column. Water clarity affects the depth to which sufficient light can be transmitted. In ideal conditions, giant kelp (*Macrocystis* spp.) can grow as much as 30–60 cm vertically per day. Some species, such as *Nereocystis*, are annuals, while others such as *Eisenia* are perennials, living for more than 20 years. In perennial kelp forests, maximum growth rates occur during upwelling months (typically spring and summer) and die-backs correspond to reduced nutrient availability, shorter photoperiods, and increased storm frequency.

Kelps are primarily associated with temperate and arctic waters worldwide. Of the more dominant genera, *Laminaria* is mainly associated with both sides of the Atlantic Ocean and the coasts of China and Japan; *Ecklonia* is found in Australia, New Zealand, and South Africa; and *Macrocystis* occurs throughout the northeastern and southeastern Pacific Ocean, Southern Ocean archipelagos, and in patches around Australia, New Zealand, and South Africa. The region with the greatest diversity of kelps (>20 species) is the northeastern Pacific, from north of San Francisco, California, to the Aleutian Islands, Alaska.

Although kelp forests are unknown in tropical surface waters, a few species of *Laminaria* have been known to occur exclusively in tropical deep waters. This general absence of kelp from the tropics is believed to be mostly due to insufficient nutrient levels associated with warm, oligotrophic waters. One recent study spatially overlaid the requisite physical parameters for kelp with mean oceanographic conditions has produced a model predicting the existence of subsurface kelps throughout the tropics worldwide to depths of 200 m. For a hotspot in the Galapagos Islands, the local model was improved with fine-scale data and tested; the research team found thriving kelp forests in all eight of their sampled sites, all of which had been predicted by the model, thus validated their approach. This suggests that their global model might actually be fairly accurate, and if so, kelp forests would be prolific in tropical subsurface waters worldwide. The importance of this contribution has been rapidly acknowledged within the scientific community and has prompted an entirely new trajectory of kelp forest research, particularly emphasizing the potential for a spatial refuge from climate change also the explanations to evolutionary patterns of kelps worldwide.

Ecosystem Architecture

Rockfish swimming around giant kelp

A diver in a kelp forest off the coast of California

A kelp forest off of the coast of Anacapa Island, California

Giant kelp uses gas-filled floats to keep the plant suspended, allowing the kelp blades near the ocean surface to capture light for photosynthesis

The architecture of a kelp forest ecosystem is based on its physical structure, which influences the associated species that define its community structure. Structurally, the ecosystem includes three guilds of kelp and two guilds occupied by other algae:

- Canopy kelps include the largest species and often constitute floating canopies that extend to the ocean surface (e.g., *Macrocystis* and *Alaria*).
- Stipitate kelps generally extend a few meters above the sea floor and can grow in dense aggregations (e.g., *Eisenia* and *Ecklonia*).
- Prostrate kelps lie near and along the sea floor (e.g., *Laminaria*).
- The benthic assemblage is composed of other algal species (e.g., filamentous and foliose functional groups, articulated corallines) and sessile organisms along the ocean bottom.
- Encrusting coralline algae directly and often extensively cover geologic substrate.

Multiple kelp species often co-exist within a forest; the term understory canopy refers to the stipitate and prostrate kelps. For example, a *Macrocystis* canopy may extend many meters above the seafloor towards the ocean surface, while an understory of the kelps *Eisenia* and *Pterygophora* reaches upward only a few meters. Beneath these kelps, a benthic assemblage of foliose red algae may occur. The dense vertical infrastructure with overlying canopy forms a system of microenvironments similar to those observed in a terrestrial forest, with a sunny canopy region, a partially shaded middle, and darkened seafloor. Each guild has associated organisms, which vary in their levels of dependence on the habitat, and the assemblage of these organisms can vary with kelp morphologies. For example, in California, *Macrocystis pyrifera* forests, the nudibranch *Melibe leonina*, and skeleton shrimp *Caprella californica* are closely associated with surface canopies; the kelp perch *Brachyistius frenatus*, rockfish *Sebastes* spp., and many other fishes are found within the stipitate understory; brittle stars and turban snails *Tegula* spp. are closely associated with the kelp holdfast, while various herbivores, such as sea urchins and abalone, live under the prostrate canopy; many seastars, hydroids, and benthic fishes live among the benthic assemblages; solitary corals, various gastropods, and echinoderms live over the encrusting coralline algae. In addition, pelagic fishes and marine mammals are loosely associated with kelp forests, usually interacting near the edges as they visit to feed on resident organisms.

Trophic Ecology

Sea urchins like this purple sea urchin can damage kelp forests by chewing through kelp holdfasts

The sea otter is an important predator of sea urchins

Classic studies in kelp forest ecology have largely focused on trophic interactions (the relationships between organisms and their food webs), particularly the understanding and top-down trophic processes. Bottom-up processes are generally driven by the abiotic conditions required for primary producers to grow, such as availability of light and nutrients, and the subsequent transfer of energy to consumers at higher trophic levels. For example, the occurrence of kelp is frequently correlated with oceanographic upwelling zones, which provide unusually high concentrations of nutrients to the local environment. This allows kelp to grow and subsequently support herbivores, which in turn support consumers at higher trophic levels. By contrast, in top-down processes, predators limit the biomass of species at lower trophic levels through consumption. In the absence of predation, these lower-level species flourish because resources that support their energetic requirements are not limiting. In a well-studied example from Alaskan kelp forests, sea otters (*Enhydra lutris*) control populations of herbivorous sea urchins through predation. When sea otters are removed from the ecosystem (for example, by human exploitation), urchin populations are released from predatory control and grow dramatically. This leads to increased herbivore pressure on local kelp stands. Deterioration of the kelp itself results in the loss of physical ecosystem structure and subsequently, the loss of other species associated with this habitat. In Alaskan kelp forest ecosystems, sea otters are the keystone species that mediates this trophic cascade. In Southern California, kelp forests persist without sea otters and the control of herbivorous urchins is instead mediated by a suite of predators including lobsters and large fishes, such as the California sheephead. The effect of removing one predatory species in this system differs from Alaska because redundancy exists in the trophic levels and other predatory species can continue to regulate urchins. However, the removal of multiple predators can effectively release urchins from predator pressure and allow the system to follow trajectories towards kelp forest degradation. Similar examples exist in Nova Scotia, South Africa, Australia and Chile. The relative importance of top-down versus bottom-up control in kelp forest ecosystems and the strengths of trophic interactions continue to be the subject of considerable scientific investigation.

The jeweled top snail Calliostoma annulatum grazing on a blade of giant kelp

The transition from macroalgal (i.e. kelp forest) to denuded landscapes dominated by sea urchins (or 'urchin barrens') is a widespread phenomenon, often resulting from trophic cascades like those described above; the two phases are regarded as alternative stable states of the ecosystem. The recovery of kelp forests from barren states has been documented following dramatic perturbations, such as urchin disease or large shifts in thermal conditions. Recovery from intermediate states of deterioration is less predictable and depends on a combination of abiotic factors and biotic interactions in each case.

Though urchins are usually the dominant herbivores, others with significant interaction strengths include seastars, isopods, kelp crabs, and herbivorous fishes. In many cases, these organisms feed on kelp that has been dislodged from substrate and drifts near the ocean floor rather than expend energy searching for intact thalli on which to feed. When sufficient drift kelp is available, herbivorous grazers do not exert pressure on attached plants; when drift subsidies are unavailable, grazers directly impact the physical structure of the ecosystem. Many studies in Southern California have demonstrated that the availability of drift kelp specifically influences the foraging behavior of sea urchins. Drift kelp and kelp-derived particulate matter have also been important in subsidizing adjacent habitats, such as sandy beaches and the rocky intertidal.

Patch Dynamics

Another major area of kelp forest research has been directed at understanding the spatial-temporal patterns of kelp patches. Not only do such dynamics affect the physical landscape, but they also affect species that associate with kelp for refuge or foraging activities. Large-scale environmental disturbances have offered important insights concerning mechanisms and ecosystem resilience. Examples of environmental disturbances include:

- Acute and chronic pollution events have been shown to impact southern California kelp forests, though the intensity of the impact seems to depend on both the nature of the contaminants and duration of exposure. Pollution can include sediment deposition and eutrophication from sewage, industrial byproducts and contaminants like PCBs and heavy metals (for example, copper, zinc), runoff of organophosphates from agricultural areas, anti-fouling chemicals used in harbors and marinas (for example, TBT and creosote) and land-based pathogens like fecal coliform bacteria.
- Catastrophic storms can remove surface kelp canopies through wave activity, but usually leave understory kelps intact; they can also remove urchins when little spatial refuge is available. Interspersed canopy clearings create a seascape mosaic where sunlight penetrates deeper into the kelp forest and species that are normally light-limited in the understory can flourish. Similarly, substrate cleared of kelp holdfasts can provide space for other sessile species to establish themselves and occupy the seafloor, sometimes directly competing with juvenile kelp and even inhibiting their settlement.
- El Niño-Southern Oscillation (ENSO) events involve the depression of oceanographic thermoclines, severe reductions of nutrient input, and changes in storm patterns. Stress due to warm water and nutrient depletion can increase the susceptibility of kelp to storm damage and herbivorous grazing, sometimes even prompting phase shifts to urchin-dominated landscapes. In general, oceanographic conditions (that is, water temperature, currents) influence the recruitment success of kelp and its competitors, which clearly affect subsequent species interactions and kelp forest dynamics.

- Overfishing higher trophic levels that naturally regulate herbivore populations is also recognized as an important stressor in kelp forests. As described in the previous section, the drivers and outcomes of trophic cascades are important for understanding spatial-temporal patterns of kelp forests.

In addition to ecological monitoring of kelp forests before, during, and after such disturbances, scientists try to tease apart the intricacies of kelp forest dynamics using experimental manipulations. By working on smaller spatial-temporal scales, they can control for the presence or absence of specific biotic and abiotic factors to discover the operative mechanisms. For example, in southern Australia, manipulations of kelp canopy types demonstrated that the relative amount of *Ecklonia radiata* in a canopy could be used to predict understory species assemblages; consequently, the proportion of *E. radiata* can be used as an indicator of other species occurring in the environment.

Human use

Kelp forests have been important to human existence for thousands of years. Indeed, many now theorise that the first colonisation of the Americas was due to fishing communities following the Pacific kelp forests during the last ice age. One theory contends that the kelp forests that would have stretched from northeast Asia to the American Pacific coast would have provided many benefits to ancient boaters. The kelp forests would have provided many sustenance opportunities, as well as acting as a type of buffer from rough water. Besides these benefits, researchers believe that the kelp forests might have helped early boaters navigate, acting as a type of "kelp highway". Theorists also suggest that the kelp forests would have helped these ancient colonists by providing a stable way of life and preventing them from having to adapt to new ecosystems and develop new survival methods even as they traveled thousands of miles. Modern economies are based on fisheries of kelp-associated species such as lobster and rockfish. Humans can also harvest kelp directly to feed aquaculture species such as abalone and to extract the compound alginic acid, which is used in products like toothpaste and antacids. Kelp forests are valued for recreational activities such as SCUBA diving and kayaking; the industries that support these sports represent one benefit related to the ecosystem and the enjoyment derived from these activities represents another. All of these are examples of ecosystem services provided specifically by kelp forests.

Threats and Management

Given the complexity of kelp forests – their variable structure, geography, and interactions – they pose a considerable challenge to environmental managers. Extrapolating even well-studied trends to the future is difficult because interactions within the ecosystem will change under variable conditions, not all relationships in the ecosystem are understood, and the nonlinear thresholds to transitions are not yet recognized. With respect to kelp forests, major issues of concern include marine pollution and water quality, kelp harvesting and fisheries, invasive species, and climate change. The most pressing threat to kelp forest preservation may be the overfishing of coastal ecosystems, which by removing higher trophic levels facilitates their shift to depauperate urchin barrens. The maintenance of biodiversity is recognized as a way of generally stabilizing ecosystems and their services through mechanisms such as functional compensation and reduced susceptibility to foreign species invasions.

The nudibranch Melibe leonina on a Macrocystis frond (California): Marine protected areas are one way to guard kelp forests as an ecosystem

In many places, managers have opted to regulate the harvest of kelp and the taking of kelp forest species by fisheries. While these may be effective in one sense, they do not necessarily protect the entirety of the ecosystem. Marine protected areas (MPAs) offer a unique solution that encompasses not only target species for harvesting, but also the interactions surrounding them and the local environment as a whole. Direct benefits of MPAs to fisheries (for example, spillover effects) have been well documented around the world. Indirect benefits have also been shown for several cases among species such as abalone and fishes in Central California. Most importantly, MPAs can be effective at protecting existing kelp forest ecosystems and may also allow for the regeneration of those that have been affected.

Culture of Microalgae in Hatcheries

Microalgae or microscopic algae grow in either marine or freshwater systems. They are primary producers in the oceans that convert water and carbon dioxide to biomass and oxygen in the presence of sunlight.

The oldest documented use of microalgae was 2000 years ago, when the Chinese used the cyanobacteria *Nostoc* as a food source during a famine. Another type of microalgae, the cyanobacteria *Arthrospira* (Spirulina), was a common food source among populations in Chad and Aztecs in Mexico as far back as the 16th century.

Today cultured microalgae is used as direct feed for humans and land-based farm animals, and as feed for cultured aquatic species such as molluscs and the early larval stages of fish and crustaceans. It is a potential candidate for biofuel production. Microalgae can grow 20 or 30 times faster than traditional food crops, and has no need to compete for arable land. Since microalgal production is central to so many commercial applications, there is a need for production techniques which increase productivity and are economically profitable.

Commonly Cultivated Microalgae Species

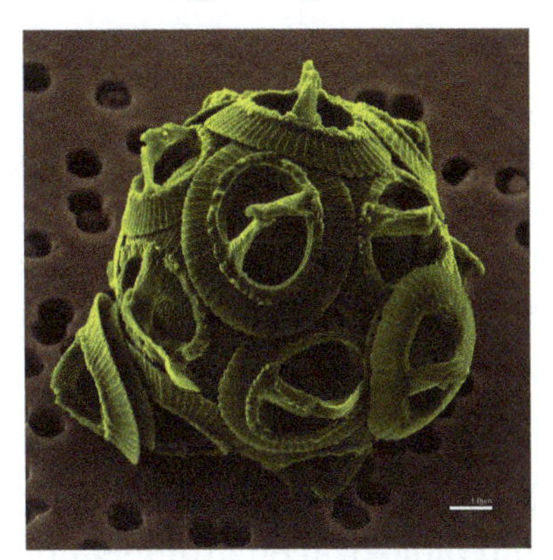

Microalgae are microscopic forms of algae, like this coccolithophore which are between 5 and 100 micrometres across

Species	Application
Chaetoceros sp.	Aquaculture
Chlorella vulgaris	Source of natural antioxidants
Dunaliella salina	Produce carotenoids (β-carotene)
Haematococcus sp.	Produce carotenoids (β-carotene), astaxanthin, canthaxanthin
Phaeodactylum tricornutum	Source of antioxidants
Porphyridium cruentum	Source of antioxidants
Rhodella sp.	Colourant for cosmetics
Skeletonema sp	Aquaculture
Arthrospira maxima	High protein content – Nutritional supplement
Arthrospira platensis	High protein content – Nutritional supplement

Hatchery Production Techniques

A range of microalgae species are produced in hatcheries and are used in a variety of ways for commercial purposes. Studies have estimated main factors in the success of a microalgae hatchery system as the dimensions of the container/bioreactor where microalgae is cultured, exposure to light/irradiation and concentration of cells within the reactor.

Open Pond System

This method has been employed since the 1950s. There are two main advantages of culturing microalgae using the open pond system. Firstly, an open pond system is easier to build and operate. Secondly, open ponds are cheaper than closed bioreactors because closed bioreactors require a cooling system. However, a downside to using open pond systems is decreased productivity of certain commercially important strains such as *Arthrospira sp.*, where optimal growth is limited by temperature. That said, it is possible to use waste heat and CO_2 from industrial sources to compensate this.

Air-lift Method

This method is used in outdoor cultivation and production of microalgae; where air is moved within a system in order to circulate water where microalgae is growing. The culture is grown in transparent tubes that lie horizontally on the ground and are connected by a network of pipes. Air is passed through the tube such that air escapes from the end that rests inside the reactor that contains the culture and creates an effect like stirring.

Closed Reactors

The biggest advantage of culturing microalgae within a closed system provides control over the physical, chemical and biological environment of the culture. This means factors that are difficult to control in open pond systems such as evaporation, temperature gradients and protection from ambient contamination make closed reactors favoured over open systems. Photobioreactors are the primary example of a closed system where abiotic factors can be controlled for. Several closed systems have been tested to date for the purposes of culturing microalgae, few important ones are mentioned below:

Horizontal Photobioreactors

This system includes tubes laid on the ground to form a network of loops. Mixing of microalgal suspended culture occurs through a pump that raises the culture vertically at timed intervals into a photobioreactor. Studies have found pulsed mixing at intervals produces better results than the use of continuous mixing. Photobioreactors have also been associated with better production than open pond systems as they can maintain better temperature gradients. An example noted in higher production of *Arthrospira sp.* used as a dietary supplement was attributed to higher productivity because of a better suited temperature range and an extended cultivation period over summer months.

Vertical Systems

These reactors use vertical polyethylene sleeves hung from an iron frame. Glass tubes can also be used alternatively. Microalgae are also cultured in vertical alveolar panels (VAP) that are a type of photobioreactor. This photobioreactor is characterised by low productivity. However, this problem can be overcome by modifying the surface area to volume ratio; where a higher ratio can increase productivity. Mixing and deoxygenation are drawbacks of this system and can be addressed by bubbling air continuously at a mean flow rate. The two main types of vertical photobioreactors are the Flow-through VAP and the Bubble Column VAP.

Flat Plate Reactors

Flat plate reactors(FPR) are built using narrow panels and are placed horizontally to maximise sunlight input to the system. The concept behind FPR is to increase the surface area to volume ratio such that sunlight is efficiently used. This system of microalgae culture was originally thought to be expensive and incapable of circulating the culture. Therefore, FPRs were considered to be unfeasible overall for the commercial production of microalgae. However, an experimental FPR system in the 1980s used circulation within the culture from a gas exchange unit across horizontal panels. This overcomes issues of circulation and provides an advantage of an open gas transfer unit that reduces oxygen build up. Examples of successful use of FPRs can be seen in the production of *Nannochloropsis sp.* used for its high levels of astaxanthin.

Fermentor-type Reactors

Fermentor-type reactors (FTR) are bioreactors where fermentation is carried out. FTRs have not developed hugely in the cultivation of microalgae and pose a disadvantage in the surface area to volume ratio and a decreased efficiency in utilizing sunlight. FTR have been developed using a combination of sun and artificial light have led to lowering production costs. However, information available on large scale counterparts to the laboratory-scale systems being developed is very limited. The main advantage is that extrinsic factors i.e. light can be controlled for and productivity can be enhanced so that FTR can become an alternative for products for the pharmaceutical industry.

Commercial Applications

Aquaculture

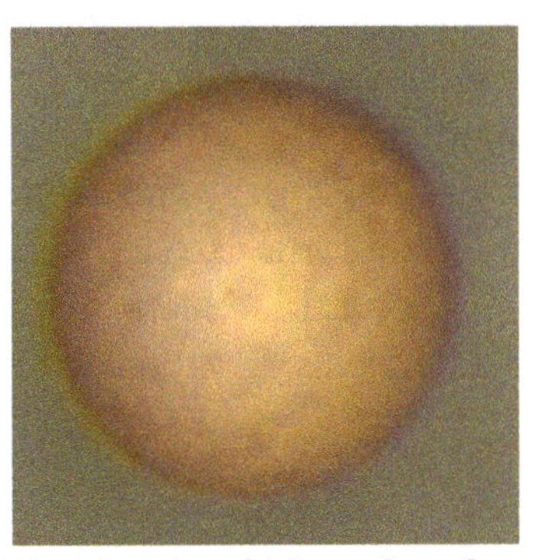

Microalgae is used to culture brine shrimp, which produce dormant eggs (pictured). The eggs can then be hatched on demand and feed to cultured fish larvae and crustaceans

Microalgae is an important source of nutrition and is used widely in the aquaculture of other organisms, either directly or as an added source of basic nutrients. Aquaculture farms rearing larvae of molluscs, echinoderms, crustaceans and fish use microalgae as a source of nutrition. Low bacteria and high microalgal biomass is a crucial food source for shellfish aquaculture.

Microalgae can form the start of a chain of further aquaculture processes. For example, microalgae is an important food source in the aquaculture of brine shrimp. Brine shrimp produce dormant eggs, called cysts, which can be stored for long periods and then hatched on demand to provide a convenient form of live feed for the aquaculture of larval fish and crustaceans.

Other applications of microalgae within aquaculture include increasing the aesthetic appeal of finfish bred in captivity. One such example can be noted in the aquaculture of salmon, where microalgae is used to make salmon flesh pinker. This is achieved by the addition of natural pigments containing carotenoids such as astaxanthin produced from the microalgae *Haematococcus* to the diet of farmed animals.

Biofuel Production

In order to meet the demands of fossil fuels, alternate means of fuels are being investigated. Biodiesel and bioethanol are renewable fuels with much potential that are important in current research. However, agriculture based renewable fuels may not be completely sustainable and thus may not be able to replace fossil fuels. Microalgae can be remarkably rich in oils (up to 80% dry weight of biomass) suitable for conversion to fuel. Furthermore, microalgae are more productive than land based agricultural crops and could therefore be more sustainable in the long run. Microalgae for biofuel production is mainly produced using tubular photobioreactors.

Cosmetic and Health Benefits

The main species of microalgae grown as health foods are *Chlorella sp.* and *Spirulina sp.* The main forms of production occur in small scale ponds with artificial mixers. Novel bioactive chemical compounds can be isolated from microalgae like sulphated polysaccharides. These compounds include fucoidans, carrageenans and ulvans that are used for their beneficial properties. These properties are anticoagulants, antioxidants, anticancer agents that are being tested in research. Red microalgae are characterised by pigments called phycobiliproteins that contain natural colourants used in pharmaceuticals and cosmetics. Production of long chain omega-3 polyunsaturated fatty acids important for human diet can also be cultured through microalgal hatchery systems.

Biofertilizer

Blue green alga was first used as a means of fixing nitrogen by allowing cyanobacteria to multiply in the soil. Nitrogen fixation is important as a means of allowing inorganic compounds such as nitrogen to be converted to organic forms which can then be used by plants. The use of cyanobacteria is an economically sound and environmentally friendly method of increasing productivity. Rice production in India and Iran have employed this method of using the nitrogen fixing properties of free living cyanobacteria to supplement nitrogen content in soils.

Other uses

Microalgae are a source of valuable molecules such as isotopes i.e. chemical variants of an element that contain different neutrons. Microalgae can effectively incorporate isotopes of carbon (^{13}C), nitrogen (^{15}N) and hydrogen (^{2}H) into their biomass. ^{13}C and ^{15}N are used to track the flow of carbon between different trophic levels/food webs. Carbon, nitrogen and sulphur isotopes can also be used to determine disturbances to bottom dwelling communities that are otherwise difficult to study.

Issues

Cell fragility is the biggest issue that limits the productivity from closed photobioreactors. Damage to cells can be attributed to the turbulent flow within the bioreactor which is required to create mixing so light is available to all cells.

Photobioreactor

A photobioreactor (PBR) is a bioreactor that utilizes a light source to cultivate phototrophic microorganisms. These organisms use photosynthesis to generate biomass from light and carbon dioxide and include plants, mosses, macroalgae, microalgae, cyanobacteria and purple bacteria. Within the artificial environment of a photobioreactor, specific conditions are carefully controlled for respective species. Thus, a photobioreactor allows much higher growth rates and purity levels than anywhere in nature or habitats similar to nature. Hypothetically, phototropic biomass could be derived from nutrient-rich wastewater and flue gas carbon dioxide in a photobioreactor.

Open Systems

The first approach for the controlled production of phototrophic organisms was a natural open pond or artificial raceway pond. Therein, the culture suspension, which contains all necessary nutrients and carbon dioxide, is pumped around in a cycle, being directly illuminated from sunlight via the liquid's surface. This construction principle is the simplest way of production for phototrophic organisms. But due to their depth (up to 0.3 m) and the related reduced average light supply, open systems only reach limited areal productivity rates. In addition, the consumption of energy is relatively high, as high amounts of water containing low product concentration have to be processed. Open space is expensive in areas with a dense population, while water is rare in others. Using open technologies causes high losses of water due to evaporation into the atmosphere.

Closed Systems

Since the 1950s several approaches have been conducted to develop closed systems, which theoretically provide higher cell densities of phototrophic organisms and therefore a lower demand of water to be pumped than open systems. In addition, closed construction avoids system-related water losses and the risk of contamination through landing water birds or dust is minimized. All modern photobioreactors have tried to balance between a thin layer of culture suspension, optimized light application, low pumping energy consumption, capital expenditure and microbial purity. Many different systems have been tested, but only a few approaches were able to perform at an industrial scale.

Redesigned Laboratory Fermenters

The simplest approach is the redesign of the well-known glass fermenters, which are state of the art in many biotechnological research and production facilities worldwide. The moss reactor for example shows a standard glass vessel, which is externally supplied with light. The existing head nozzles are used for sensor installation and for gas exchange. This type is quite common in laboratory scale, but it has never been established in bigger scale, due to its limited vessel size.

Tubular Photobioreactors

Tubular glass photobioreactor

Made from glass or plastic tubes, this photobioreactor type has succeeded within production scale. The tubes are oriented horizontally or vertically and are supplied from a central utilities installation with pump, sensors, nutrients and CO_2. Tubular photobioreactors are established worldwide from laboratory up to production scale, e.g. for the production of the carotenoid Astaxanthine from the green algae *Haematococcus pluvialis* or for the production of food supplement from the green algae *Chlorella vulgaris*. These photobioreactors take advantage from the high purity levels and their efficient outputs. The biomass production can be done at a high quality level and the high biomass concentration at the end of the production allows energy efficient downstream processing. Due to the recent prices of the photobioreactors, economically feasible concepts today can only be found within high-value markets, e.g. food supplement or cosmetics.

The advantages of tubular photobioreactors at production scale are also transferred to laboratory scale. A combination of the mentioned glass vessel with a thin tube coil allows relevant biomass production rates a laboratory research scale. Being controlled by a complex process control system, the regulation of the environmental conditions reaches a high level.

Christmas Tree Photobioreactor

Christmas tree reactor

An alternative approach is shown by a photobioreactor, which is built in a tapered geometry and which carries a helically attached, translucent double hose circuit system. The result is a layout similar to a Christmas tree. The tubular system is constructed in modules and can theoretically be scaled outdoors up to agricultural scale. A dedicated location is not crucial, similar to other closed systems, and therefore non-arable land is suitable as well. The material choice should prevent biofouling and ensure high final biomass concentrations. The combination of turbulence and the closed concept should allow a clean operation and a high operational availability.

Plate Photobioreactor

Plastic plate photobioreactor

Another development approach can be seen with the construction based on plastic or glass plates. Plates with different technical design are mounted to form a small layer of culture suspension, which provides an optimized light supply. In addition, the simpler construction compared to tubular reactors allows the use of less expensive plastic materials. From the pool of different concepts e.g. meandering flow designs or bottom gassed systems have been realized and shown good output results. Some unsolved issues are material life time stability or the biofilm forming. Applications at industrial scale are limited by the scalability of plate systems.

In April 2013, the IBA in Hamburg, Germany, a building with an integrated glass plate photobioreactor facade, was commissioned.

Horizontal Photobioreactor

Horizontal photobioreactor with zigzag shaped geometry

This photobioreactor type consists of a plate-shaped basic geometry with peaks and valleys arranged in regular distance. This geometry causes the distribution of incident light over a larger surface which corresponds to a dilution effect. This also helps solving a basic problem in phototrophic cultivation, because most microalgae species react sensitively to high light intensities. Most microalgae experience light saturation already at light intensities, ranging substantially below the maximum daylight intensity of approximately 2000 W/m². Simultaneously, a larger light quantity can be exploited in order to improve photoconversion efficiency. The mixing is accomplished by a rotary pump, which causes a cylindrical rotation of the culture broth. In contrast to vertical designs, horizontal reactors contain only thin layers of media with a correspondingly low hydrodynamic pressure. This has a positive impact on the necessary energy input and reduces material costs at the same time.

Foil Photobioreactor

The pressure of market prices has led the development of foil-based photobioreactor types. Inexpensive PVC or PE foils are mounted to form bags or vessels which cover algae suspensions and expose them to light. The pricing ranges of photobioreactor types have been enlarged with the foil systems. It has to be kept in mind, that these systems have a limited sustainability as the foils have to be replaced from time to time. For full balances, the investment for required support systems has to be calculated as well.

Porous Substrate Bioreactor

Porous Substrate Bioreactor (PSBR), being developed at University of Cologne, also known as the twin-layer system, uses a new principle to separate the algae from a nutrient solution by means of a porous reactor surface on which the microalgae are trapped in biofilms. This new procedure reduces by a factor of up to one hundred the amount of liquid needed for operation compared to the current technology, which cultivates algae in suspensions. As such, the PSBR procedure significantly reduces the energy needed while increasing the portfolio of algae that can be cultivated.

Seaweed Farming

Seaweed farming is the practice of cultivating and harvesting seaweed. In its simplest form, it consists of the management of naturally found batches. In its most advanced form, it consists of fully controlling the life cycle of the algae. The main food species grown by aquaculture in Japan, China and Korea include Gelidium, Pterocladia, Porphyra, and Laminaria. Seaweed farming has frequently been developed as an alternative to improve economic conditions and to reduce fishing pressure and overexploited fisheries. Seaweeds have been harvested throughout the world as a food source as well as an export commodity for production of agar and carrageenan products.

Global production of farmed aquatic plants, overwhelmingly dominated by seaweeds, grew in output volume from 13.5 million tonnes in 1995 to just over 30 million tonnes in 2016. Cultivation of gim (laver) in Korea is reported in books from the 15th century, such as Revised and Augmented Survey of the Geography of Korea and Geography of Gyeongsang Province.

Seaweed farming began in Japan as early as 1670 in Tokyo Bay. In autumn of each year, farmers would throw bamboo branches into shallow, muddy water, where the spores of the seaweed would collect. A few weeks later these branches would be moved to a river estuary. The nutrients from the river would help the seaweed to grow.

In the 1940s, the Japanese improved this method by placing nets of synthetic material tied to bamboo poles. This effectively doubled the production. A cheaper variant of this method is called the hibi method — simple ropes stretched between bamboo poles.

Seaweed farming has been introduced to several different islands in Indonesia, such as Rote island. The Australian government first brought agar seeds to Rote 15 years ago to keep locals out of poverty. In the early 1970s, there was a recognized demand for seaweed and seaweed products, outstripping supply, and cultivation was viewed as the best means to increase productions.

Culture Methods

The earliest seaweed farming guides in the Philippines recommended the cultivation of Laminaria seaweed and reef flats at approximately one meter's depth at low tide. They also recommended cutting off seagrasses and removing sea urchins before farm construction. Seedlings are then tied to monofilament lines and strung between mangrove stakes pounded into the substrate. This off-bottom method is still one of the primary methods used today.

There are new long-line cultivation methods that can be used in deeper water approximately 7 meters in depth. They use floating cultivation lines anchored to the bottom and are the primary methods used in the villages of North Sulawesi, Indonesia.

Cultivation of seaweed in Asia is a relatively low-technology business with a high labor requirement. There have been many attempts in various countries to introduce high technology to cultivate detached plants growth in tanks on land in order to reduce labor, but they have yet to attain commercial viability.

Environmental and Ecological Impacts

Several environmental problems can result from seaweed farming. Sometimes seaweed farmers cut down mangroves to use as stakes for their ropes. This, however, negatively affects farming since it reduces the water quality and mangrove biodiversity due to depletion. Farmers may also sometimes remove eelgrass from their farming areas. This, however, is also discouraged, as it adversely affects water quality.

Seaweed farming helps to preserve coral reefs by increasing diversity where the algae and seaweed have been introduced, and it also provides an added niche for local species of fish and invertebrates. Farming may be beneficial by increasing the production of herbivorous fishes and shellfish in the area. Pollnac & et al 1997b reported an increase in Siginid population after the start of extensive farming of eucheuma seaweed in villages in North Sulawesi, Indonesia.

Seaweed culture can also be used to capture, absorb, and eventually incorporate excessive nutrients into living tissue. "Nutrient bioextraction" is the preferred term for bioremediation involving cultured plants and animals. Nutrient bioextraction (also called bioharvesting) is the practice of farming and harvesting shellfish and seaweed to remove nitrogen and other nutrients from natural water bodies. Seaweed farming can be an actor in biological carbon sequestration.

Societal Impact

The practice of seaweed farming has long since spread beyond Japan. In 1997, it was estimated that 40,000 people in the Philippines made their living through seaweed farming. Cultivation is also common in all of southeast Asia, Canada, Great Britain, Spain, and the United States.

Harvesting seaweed in North Cape (Canada)

Socioeconomic Aspects

In Japan alone, the annual production value of nori amounts to US$2 billion and is one of the world's most valuable crops produced by aquaculture. The high demand for seaweed production provides plentiful opportunities and work for the local community. A study conducted by the Philippines showed that plots of approximately one hectare could have a net income from eucheuma farming that was 5 to 6 times that of the minimum average wage of an agriculture worker. In the same study, they also saw an increase in seaweed exports from 675 metric tons (MT) in 1967 to 13,191 MT in 1980, which doubled to 28,000 MT by 1988.

Algae Bioreactor

An algae bioreactor is used for cultivating micro or macro algae. Algae may be cultivated for the purposes of biomass production (as in a seaweed cultivator), wastewater treatment, CO_2 fixation, or aquarium/pond filtration in the form of an algae scrubber. Algae bioreactors vary widely in design, falling broadly into two categories: open reactors and enclosed reactors. Open reactors are exposed to the atmosphere while enclosed reactors, also commonly called photobioreactors, are isolated to varying extent from the atmosphere. Specifically, algae bioreactors can be used to produce fuels such as biodiesel and bioethanol, to generate animal feed, or to reduce pollutants such as NO_x and CO_2 in flue gases of power plants. Fundamentally, this kind of bioreactor is based on the photosynthetic reaction which is performed by the chlorophyll-containing algae itself using dissolved carbon dioxide and sunlight energy. The carbon dioxide is dispersed into the reactor fluid to make it accessible for the algae. The bioreactor has to be made out of transparent material.

The algae are photoautotroph organisms which perform oxygenic photosynthesis.

The equation for photosynthesis:

$$6\,CO_2 + 6\,H_2O \rightarrow C_6H_{12}O_6 + 6\,O_2 \qquad \Delta H^0 = +2870\,\frac{kJ}{mol}$$

Some of the first experiments with the aim of cultivating algae were conducted in 1957 by the "Carnegie Institution" in Washington. In these experiments, monocellular Chlorella were cultivated by adding CO_2 and some minerals. In the early days, bioreactors were used which were made of glass and later changed to a kind of plastic bag. The goal of all this research has been the cultivation of algae to produce a cheap animal feed.

Frequently used Photo Reactor Types

Nowadays 3 basic types of algae photobioreactors have to be differentiated, but the determining factor is the unifying parameter – the available intensity of sunlight energy.

Plate Photobioreactor

A plate reactor simply consists of vertically arranged or inclined rectangular boxes which are often divided in two parts to affect an agitation of the reactor fluid. Generally these boxes are arranged to a system by linking them. Those connections are also used for making the process of filling/emptying, introduction of gas and transport of nutritive substances, easier. The introduction of the flue gas mostly occurs at the bottom of the box to ensure that the carbon dioxide has enough time to interact with algae in the reactor fluid.

Tubular Photobioreactor

A tubular reactor consists of vertical or horizontal arranged tubes, connected together to a pipe system. The algae-suspended fluid is able to circulate in this tubing. The tubes are generally made out of transparent plastics or borosilicate glass and the constant circulation is kept up by a pump at the end of the system. The introduction of gas takes place at the end/beginning of the tube system. This way of introducing gas causes the problem of deficiency of carbon dioxide, high concentration of oxygen at the end of the unit during the circulation, and bad efficiency.

Bubble Column Photobioreactor

A bubble column photo reactor consists of vertical arranged cylindrical column, made out of transparent material. The introduction of gas takes place at the bottom of the column and causes a turbulent stream to enable an optimum gas exchange. At present these types of reactors are built with a maximum diameter of 20 cm to 30 cm in order to ensure the required supply of sunlight energy.

The biggest problem with the sunlight determined construction is the limited size of the diameter. Feuermann et al. invented a method to collect sunlight with a cone shaped collector and transfer it with some fiberglass cables which are adapted to the reactor in order to enable constructions of a column reactor with wider diameters. - on this scale the energy consumption due to pumps etc. and the CO_2 cost of manufacture may outweigh the CO_2 captured by the reactor.

Industrial Usage

The cultivation of algae in a photobioreactor creates a narrow range of industrial application possibilities. Some power companies already established research facilities with algae photobioreactors to find out how efficient they could be in reducing CO_2 emissions, which are contained in flue gas, and how much biomass will be produced. Algae biomass has many uses and can be sold to generate additional income. The saved emission volume can bring an income too, by selling emission credits to other power companies.

The utilisation of algae as food is very common in East Asian regions. Most of the species contain only a fraction of usable proteins and carbohydrates, and a lot of minerals and trace elements. Generally, the consumption of algae should be minimal because of the high iodine content, particularly problematic for those with hyperthyroidism. Likewise, many species of diatomaceous algae produce compounds unsafe for humans. The algae, especially some species which contain over 50 percent oil and a lot of carbohydrates, can be used for producing biodiesel and bioethanol by extracting and refining the fractions. This point is very interesting, because the algae biomass is generated 30 times faster than some agricultural biomass, which is commonly used for producing biodiesel.

Algae Harvesting

Algae harvesting is the process of cultivating algae for use as biofuel. An alternative to traditional fossil fuels, biofuels have a lower environmental impact. Grown in saltwater, they have a high flash point, and they are biodegradable. Not to mention, they release no new CO_2 into the atmosphere during production. For this reason, algae harvesting has become an important source of plant-based fuels.

First harvested for fuel in the mid-1940's, it wasn't until the oil embargo of the 1970's that production started in earnest. The U.S. Department of Energy launched the Aquatic Species Program in 1978 hoping to develop an alternative to petroleum derived fuels. Unfortunately, oil prices decreased in the 1990's and biofuel production could no longer compete with dropping petroleum prices. The Department of Energy abandoned the program in 1996. When oil prices rose again in the early 2000's, funding returned, and a new wave of algae-based fuels entered the market. Algenol Biofuels, Sapphire Energy, and Genifuel are just a few of the companies interested in algae harvesting and biofuel production. Using patented technologies to prime algae for commercial use, these companies hope to displace hazardous fossil fuels with a viable (and competitive) plant-based alternative.

Process

To cultivate algae for biofuel use, first grow it as a crop and then process it as a fuel. Fortunately, algae require less fresh water to produce, less time to harvest, and less land to cultivate than traditional food crops. Unfortunately, it is more costly to produce than other fuels. For this reason, most algae companies focus on cost reduction and commercial viability. Many believe that with the former comes the latter.

Step One: Cultivation

The first step is growing algae. For commercial vendors, this typically involves the use of open ponds or algal turf scrubber (ATS) systems. Open ponds are long bodies of water lined with geomembrane material and cultivated with algae. Submerged in shallow water, algae absorb light from the sun and turn it into biomass. Growers collect harvests every two days. A modern alternative, ATS systems mimic coral reefs by covering a slanted surface with geomembrane material and cultivating algae on the surface. This method produces larger yields and results in less contamination than traditional open pond structures. Not to mention, ATS systems create algae with lower lipid levels making it an ideal source for many fuel options. Butanol, ethanol, and methane are just a few of the fuels produced by this method.

Step Two: Fuel Production

After harvest, dehydration of the algae takes place. This process is the most time-consuming and costly part of production. Algae has a high water content that producers remove via various extraction methods. Energy-rich compounds known as lipids are separated from the dehydrated material to create biofuel products. Biodiesel, methane, ethanol, green diesel, and jet fuels are made from these lipids, while the remaining dehydrated products are used to produce bioethanol or butanol. Though algae provide a safe, more eco-friendly alternative to traditional fossil fuels, it is still not as cost-effective to produce as petroleum based products. For this reason, there is still much research to do as the industry expands. That said, the benefits of algae production far outweigh the consequences of competing products. Algae grows in less than ideal situations as well as on land that previously considered useless. It requires minimal fresh water to harvest, and some sources say that replacing fossil fuels with biofuels could reduce CO_2 emissions by as much as 80%. For this reason, and many others, algae harvesting is an important part of biofuel creation.

References

- Algaculture, biofuels, fuel-efficiency; howstuffworks.com, Retrieved 23 May, 2019
- Abbott, Isabella A.; Isabella, Abbott; Hollenberg, George J. (1992-08-01). Marine Algae of California. Stanford University Press. ISBN 9780804721523
- Abbott, Isabella A. (July 1996). "Ethnobotany of seaweeds: clues to uses of seaweeds". Hydrobiologia. 326-327 (1): 15–20. Doi:10.1007/bf00047782. ISSN 0018-8158
- Algae-harvesting-important: btlliners.com , Retrieved 10 June, 2019
- Bushing, William W (2000). "MACROCYSTIS PYRIFERA, GIANT BLADDER KELP". Www.starthrower.org. Retrieved 2018-07-20
- Whitton, B., and M. Potts. 2000. The ecology of Cyanobacteria: their diversity in time and space p. 506, Kluwer Academic. ISBN 978-0-7923-4735-4
- Kelp forests provide habitat for a variety of invertebrates, fish, marine mammals, and birds NOAA. Updated 11 January 2013. Retrieved 15 January 2014

PERMISSIONS

All chapters in this book are published with permission under the Creative Commons Attribution Share Alike License or equivalent. Every chapter published in this book has been scrutinized by our experts. Their significance has been extensively debated. The topics covered herein carry significant information for a comprehensive understanding. They may even be implemented as practical applications or may be referred to as a beginning point for further studies.

We would like to thank the editorial team for lending their expertise to make the book truly unique. They have played a crucial role in the development of this book. Without their invaluable contributions this book wouldn't have been possible. They have made vital efforts to compile up to date information on the varied aspects of this subject to make this book a valuable addition to the collection of many professionals and students.

This book was conceptualized with the vision of imparting up-to-date and integrated information in this field. To ensure the same, a matchless editorial board was set up. Every individual on the board went through rigorous rounds of assessment to prove their worth. After which they invested a large part of their time researching and compiling the most relevant data for our readers.

The editorial board has been involved in producing this book since its inception. They have spent rigorous hours researching and exploring the diverse topics which have resulted in the successful publishing of this book. They have passed on their knowledge of decades through this book. To expedite this challenging task, the publisher supported the team at every step. A small team of assistant editors was also appointed to further simplify the editing procedure and attain best results for the readers.

Apart from the editorial board, the designing team has also invested a significant amount of their time in understanding the subject and creating the most relevant covers. They scrutinized every image to scout for the most suitable representation of the subject and create an appropriate cover for the book.

The publishing team has been an ardent support to the editorial, designing and production team. Their endless efforts to recruit the best for this project, has resulted in the accomplishment of this book. They are a veteran in the field of academics and their pool of knowledge is as vast as their experience in printing. Their expertise and guidance has proved useful at every step. Their uncompromising quality standards have made this book an exceptional effort. Their encouragement from time to time has been an inspiration for everyone.

The publisher and the editorial board hope that this book will prove to be a valuable piece of knowledge for students, practitioners and scholars across the globe.

INDEX

P

R

S

T

V

W

Z

CPSIA information can be obtained
at www.ICGtesting.com
Printed in the USA
BVHW062205091221
623625BV00003B/79